Secret Codes & Number Games

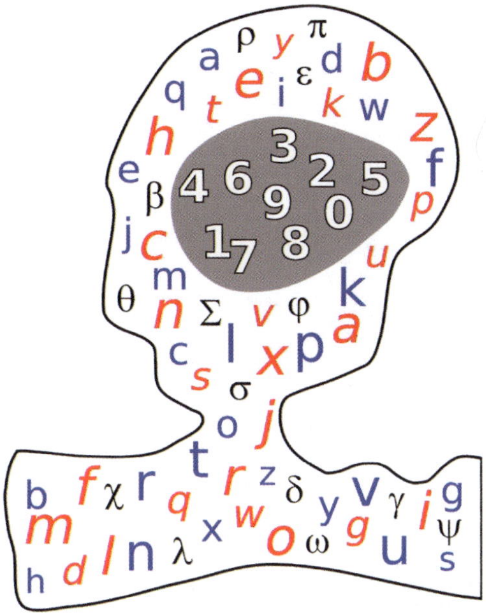

Cryptographic Projects & Number Games for Children Ages 5-16

Dev Gualtieri, Ph.D.

Tikalon Press

http://www.tikalonpress.com

ISBN: 978-1-942459-01-9

CONTENTS

AUTHOR'S FOREWORD

When I was a child, I enjoyed making and sending secret messages. Those were the days before ubiquitous computers, so everything was done using paper, pencil, and the occasional decoding ring. As you can imagine, the more complicated codes were just too much work for a child, so the Caesar cipher ruled the day, and it was more fun sending longer messages using "invisible ink."

Computers and the Internet have made cryptography much more interesting, but it's important to learn the basics and understand what's happening "under the hood" of cryptographic applications.

Advanced cryptography uses a lot of mathematics, but I've tried to shield the younger readers from most of the math. I have, however, included a few interesting number games.

ACKNOWLEDGEMENTS

My largest debt as an author is to the many other authors who have written both academic and popular accounts of cryptography over the years. I mention just a few notable examples from my own library, as follow:

- David Kahn, The Codebreakers – The Story of Secret Writing (Macmillan, 1967).
- Simon Singh, The Code Book (Doubleday, 1999).
- Steven Levy, Crypto: How the Code Rebels Beat the Government Saving Privacy in the Digital Age (Penguin Books , 2001).
- Bruce Schneier, Applied Cryptography (Wiley, 1996).

- Niels Ferguson and Bruce Schneier, Practical Cryptography (Wiley, 2003).
- Alan G. Konheim, Cryptography: A Primer (John Wiley & Sons, 1981).
- Alfred Menezes, Paul van Oorschot and Scott Vanstone, Eds., Handbook of Applied Cryptography (CRC Press, 1996).

I mention, also, that the arXiv preprint server (http://arxiv.org) has a section on Cryptography and Security. The Internet once had some very informative cryptography news groups, but these have largely disappeared. Still, many of the contributors to these groups offered many useful insights into this unusual field of applied mathematics.

There is also Wikipedia (http://en.wikipedia.org), which serves as a wealth of information on so many topics and to which I've contributed some articles of my own.

1 SECRET CODES

People have always kept secrets from each other. At home, children keep secrets from their parents, although they shouldn't. At school, and around their neighborhood, children keep secrets from other children. In the world at large, governments keep secrets from other governments, and companies keep secrets from other companies.

Keeping a secret is easy if you're the only one who knows it. It's harder if you want to tell someone else. If they're nearby, you whisper in their ear, hoping that no one else will hear you.

What if you want to tell someone a secret, but you're in your home, and the other person is in his home? Maybe you live in different cities. How do you tell someone a secret using email or postal mail so no one else will know the secret, even when they can peek and see the message?

Passing secret messages happens all the time on the Internet. When your parents use a credit card to order something online, their order is sent as a secret message. When a general needs to send a message to his troops, he might use a radio, but only his troops understand the message, since it's sent in a code.

This book will show you how secret messages are coded and decoded, and you'll learn a few ways to send secret messages, yourself.

The science of sending and receiving secret messages is called cryptography. The word, "cryptography," is the combination of two Greek words, "crypto," which means "hidden," and "graphos," which means writing.

Cryptography, as used today for things like purchasing things on the Internet, involves mathematics. Some math can be fun, so you'll learn a few number games, too.

2 SECRET CODES BEFORE COMPUTERS

In the very early days of civilization, very few people could read. If you wanted to send a mostly secret message, you would just write it down. If you spoke Latin, and you wanted to go a step further, you would use a language, such as Greek, that only a few of your friends knew.

Julius Caesar was Emperor of the Roman Empire about two thousand years ago. Before he was Emperor, he was a general. General Caesar sent secret messages to his troops using a very simple coding process, or cipher. It was a very simple cipher, but it worked well in an age when few people could read. You'll learn how to send messages using the Caesar cipher, and some of its variants, in a later chapter.

Even before Caesar, people sent secret messages by hiding them. This process is called steganography. The word "steganography," is derived from words that also mean hidden writing. Sending such a hidden message might involve coding the message first before sending it. There are two interesting examples of steganography in ancient times.

In one case, a man's head was shaved, and the message was written on his bald head. When his hair grew back, the message was hidden. It was revealed when his head was again shaved.

In another case, a message was written on a wooden board. The board was then covered with wax. Such wax tablets were a common way of writing before paper was invented. Messages could be scratched into the wax, and then erased by heating the wax. In this case, there was no message scratched into the wax, so everyone thought that the tablet was blank. Scratching off the wax revealed the real message on the board.

Closer to our own time, there were more educated people, so more of them could read and solve simple puzzles, such as the Caesar cipher. It took more effort to write secret messages (a process called encryption), and to read them (a process called decryption).

One popular method used a key word, usually called just a key. Just as for a real lock and key, a message could be locked (encrypted) using a key, and then unlocked (decrypted) using the same key. You'll learn how to use a key to write secret messages.

3 SECRET CODES, AFTER COMPUTERS

A turning point in cryptography came with the discovery of electricity and the invention of the battery, the light bulb, and the electromechanical relay. Electromechanical relays were well developed, since they were used for telephone switching, and simple mathematical functions could be done using relays in logic circuits.

A twenty-six position switch or equivalent rotary relay can be used to select a letter of the alphabet. You could have thirty-six positions if you want to use numbers, too, but since numbers can just be written as words, twenty-six is all that's really needed. Depending on how the wires are attached to the switch, selecting a letter on the switch will light a particular light representing a different letter. What makes this interesting is that the wires connected to the lights can be passed through other switches.

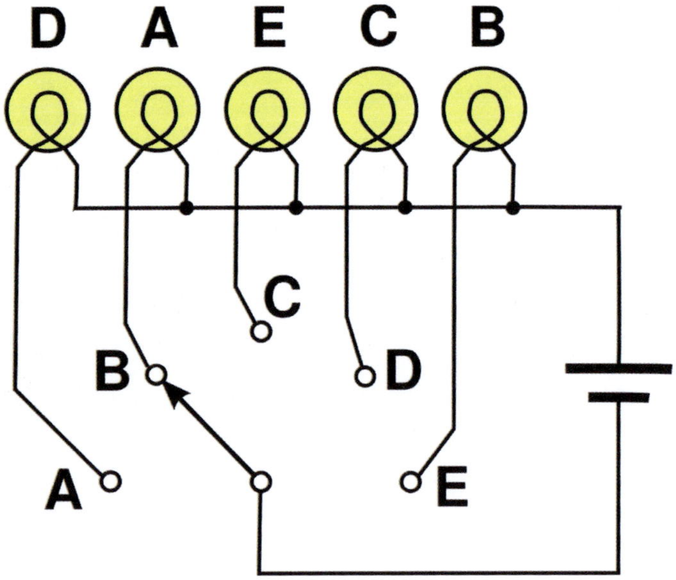

A five-position switch wired to lights and a battery.

This circuit codes the letters A-E as different letters, a process called substitution.

Wiring other switches between the battery and lights allows us to easily change the wiring selecting which bulb to light according to a rule, or algorithm. This means that each letter in a message can be replaced by not just the same letter, but different letters, depending on where it is in our message.

A particular arrangement of letters could be contained on a card with holes punched in it. These punch cars were very common in the early days of computers. The holes in the card can select which letters are transformed to other letters The codes can even be punched onto a long paper tape, so that each letter could be coded differently as the paper tape moves through the machine.

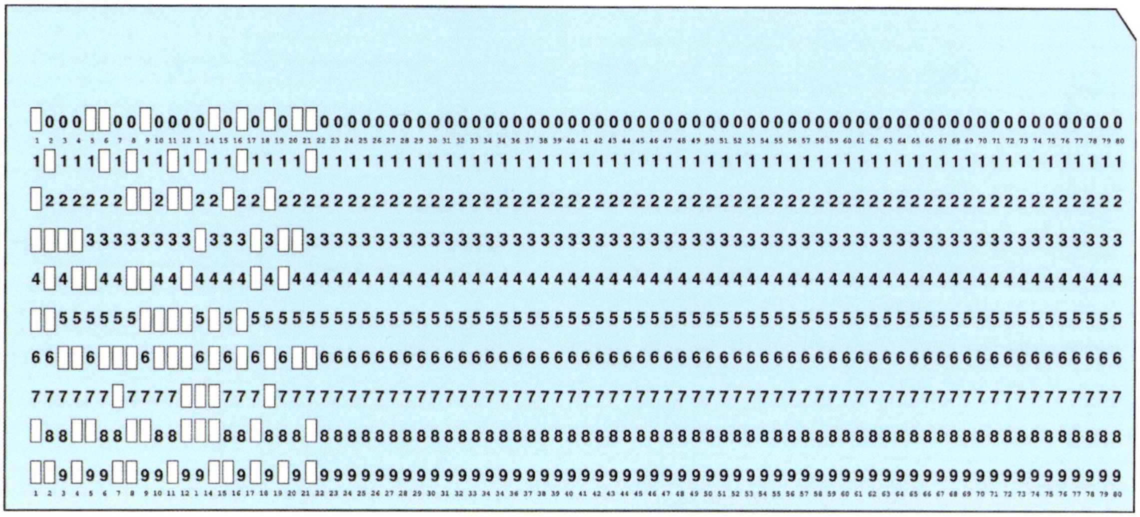

An eighty-column punch card, as used in early computers.

Such electrical circuitry allowed the creation of codes that were harder to "crack," but the real turning point came as computers became available. Instead of rotating switches, or using punched cards or paper tapes, all these operations could be done in software, the language of computers. Since computers are very fast, operations can be done on a large chunk of text, not just individual letters.

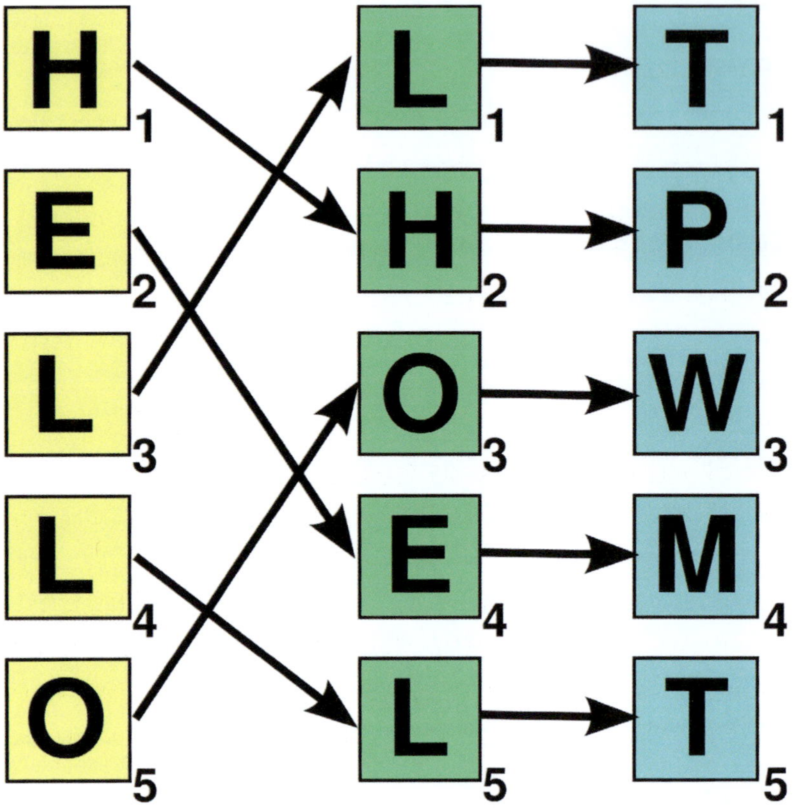

The basic cipher operations are permutation and substitution.

Permutation is changing the order of letters according to a rule.

Substitution is the replacement of each letter by another according to a rule.

Of course, the same computers used to encrypt messages can be used to crack them as well. This is especially important when the cracking computers are very fast computers called supercomputers. Even then, it's not just speed, but the knowledge of the code crackers that makes this possible. The code-cracking process is called cryptanalysis.

Cryptanalysis is important not just for cracking secret messages. It's a way to ensure that the secret codes you use can't be cracked by other people. Cryptanalysis involves a lot of mathematics, so people who do this are usually mathematicians.

A modern supercomputer. (Photograph Courtesy of Argonne National Laboratory)

4 CODING DEVICES - SCYTALE

The Scytale was a simple encryption device used thousands of years ago. It could be made from materials found in any household. A strip of paper is wrapped around a rod, as shown in the figure, and a message is written onto it, along the length of the rod, one character per paper width. When the end of the paper row is reached, you continue writing on the lower rows to complete the message. When the message is written, any remaining space is filled with random letters.

When the paper strip is unwrapped, there is no apparent message written on its length, but the message is revealed when it is wrapped on the same-sized rod. The figure shows a Scytale for the first verse of "Mary had a little lamb..." To make things easier, coded messages are usually written all in capital letters, without any spaces or punctuation.

MARYHADALITTLELAMBWHOSEFLEECEWASWHITEASSNOW

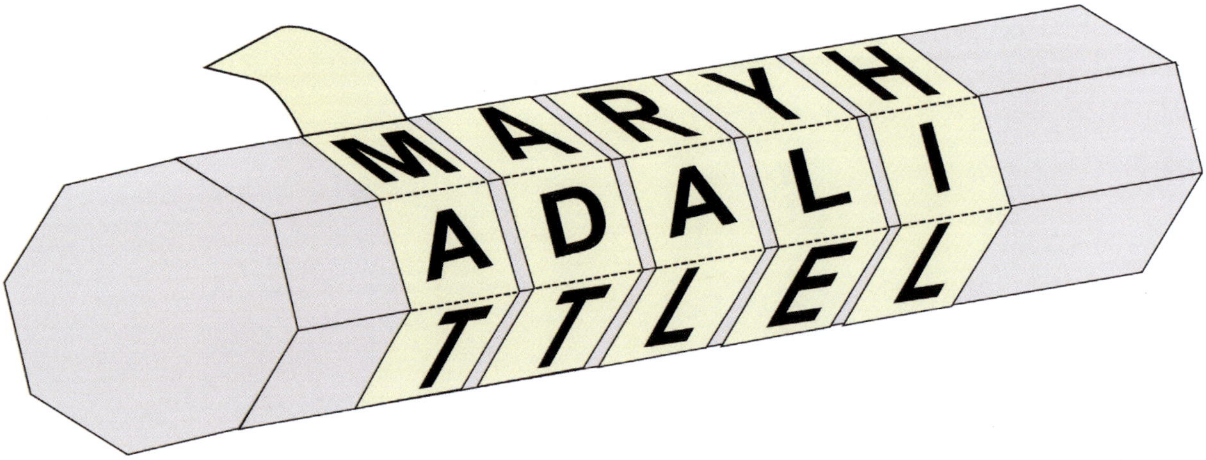

The unwrapped strip of paper for the first two verses of "Mary had a Little Lamb" is shown on the next page.

"Mary had a little lamb, whose fleece was white as snow. And everywhere that Mary went, the lamb was sure to go."

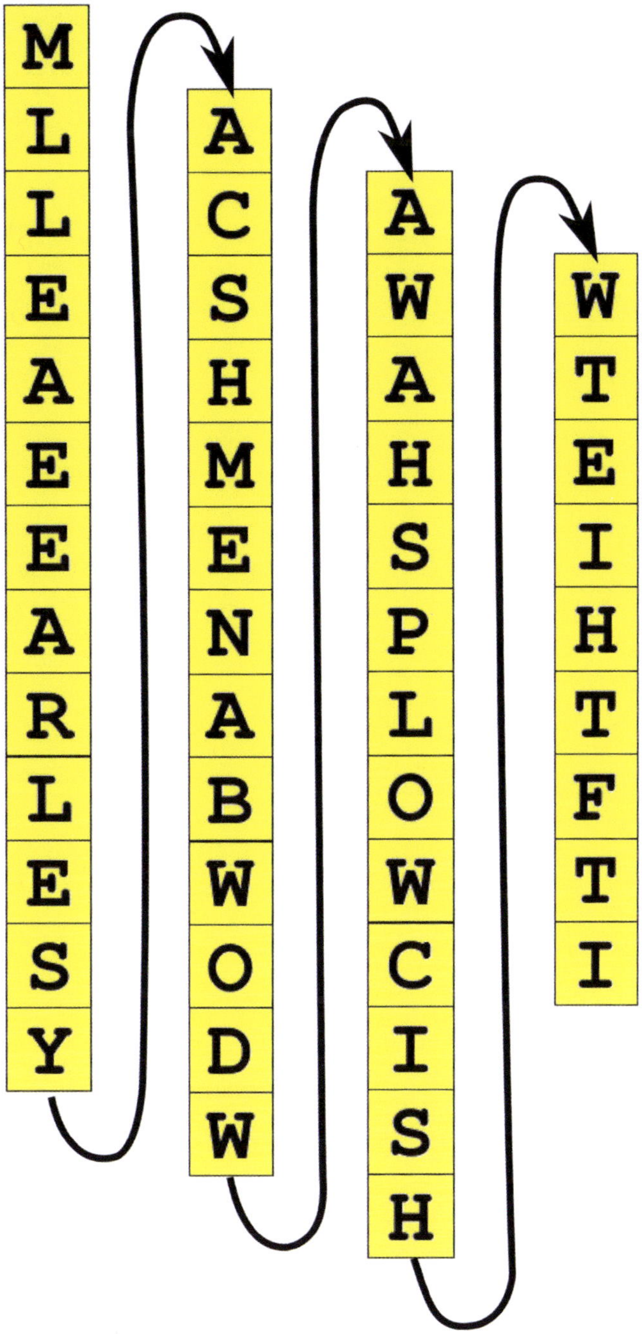

5 CODING DEVICES – DECODING WHEEL

The following page shows how you can make a coding device called a decoding wheel. The decoding wheel is also called a cipher disk, an Alberti cipher disk, or a secret decoder ring.

The decoding wheel is a mechanical aid for doing a substitution cipher. A substitute alphabet (red in the figure) is written around the edge of a disk. This disk is attached inside another disk with the alphabet in regular order (blue).

You decide on how to make the substitution by rotating the disks to match one regular character to a substitute character. In cryptography, the regular characters are called the **plaintext**, and the substitute characters are called the **ciphertext**. This is the "key" that will unlock the coded message later.

You code your message by looking at each letter in your message (blue) and writing down the substitute letter (red). The person who gets this coded message will decode it by using the key to match the letters on his decoding wheel, reading the red letters and writing the blue letters.

For example, if we choose the key where the plaintext **2** is set equal to the ciphertext **5**, then the message,

MARYHADALITTLELAMB

will be coded as

VGWQ7G9G8PSS8I8GVK

To make your own decoding wheel, have a copy of the next page made at a local copy center and paste it onto cardboard.

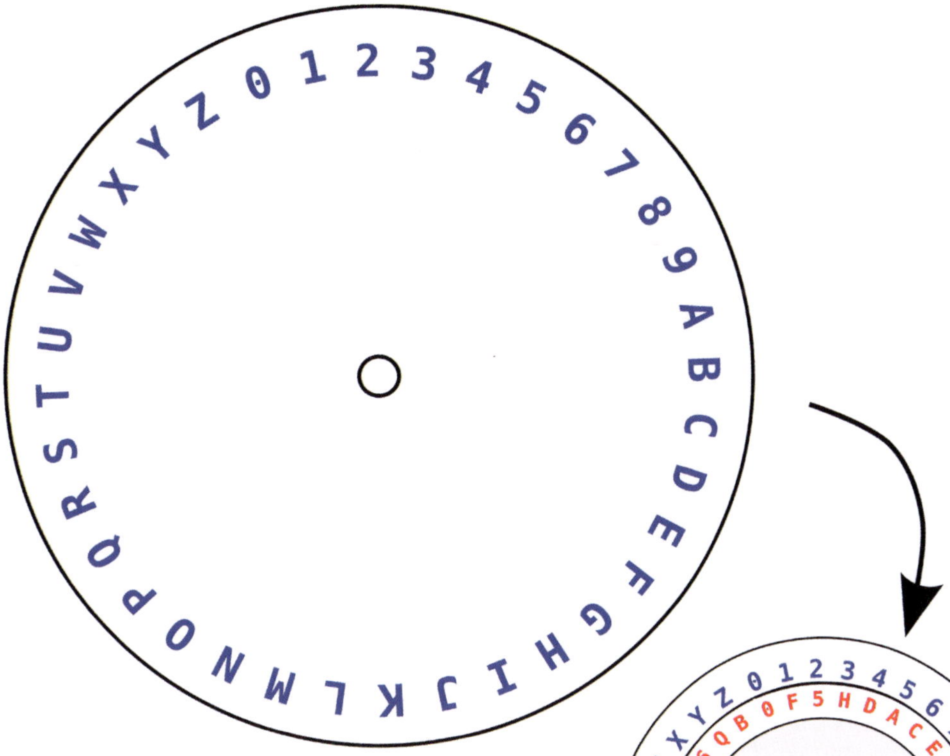

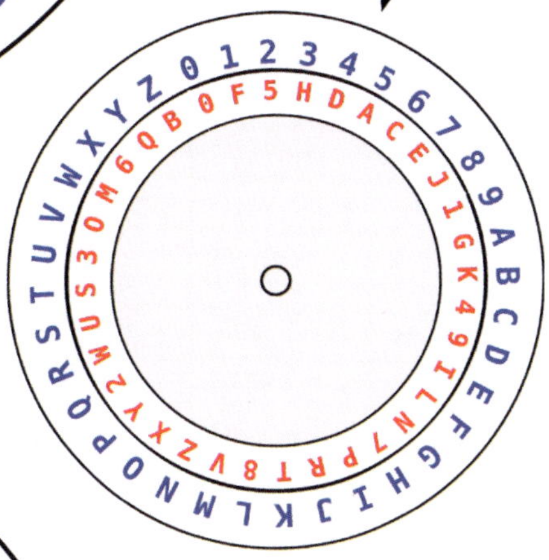

Directions: Copy this page, cut out the circles and join them together with a paper fastener at the center. Rotate to select encoding character pair.

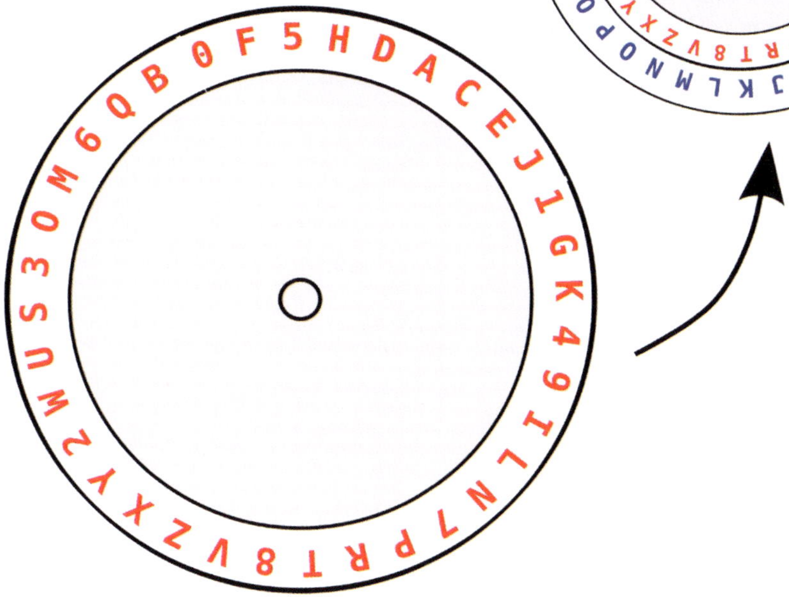

6 CODING DEVICES – CAESAR RODS

If you take the decoding wheel of the last chapter, cut it and stretch it into long strips, you would get a similar coding and decoding device. Its construction is somewhat easier, since you just need the alphabet in regular order printed or inscribed on a rod. To make it work, you need three of these rods; or, one short rod, and another rod twice as long.

Like the decoding wheel, a key is set by matching one plaintext character with a ciphertext character. This is done by sliding the rods against each other, as shown in the example on the next page, where the plaintext **A** (blue) is matched with the ciphertext **F** (red).

Since the encoding method is the same as that of the Caesar cipher presented in a later chapter, these rods are called Caesar rods. You can make and use your own set of Caesar rods by following these directions:

- Make a copy of the next page.
- Cut-out the letter columns.
- Paste them onto two pieces of wood, or thick pieces of cardboard, as shown (one piece is twice as long as the other).
- Slide the rods to select the code key (The example A=F is shown).
- Encode your message by changing your message (plaintext) letters (blue) to ciphertext letters (red).
- Decode your message by changing the ciphertext letters (red) to plaintext letters (blue).

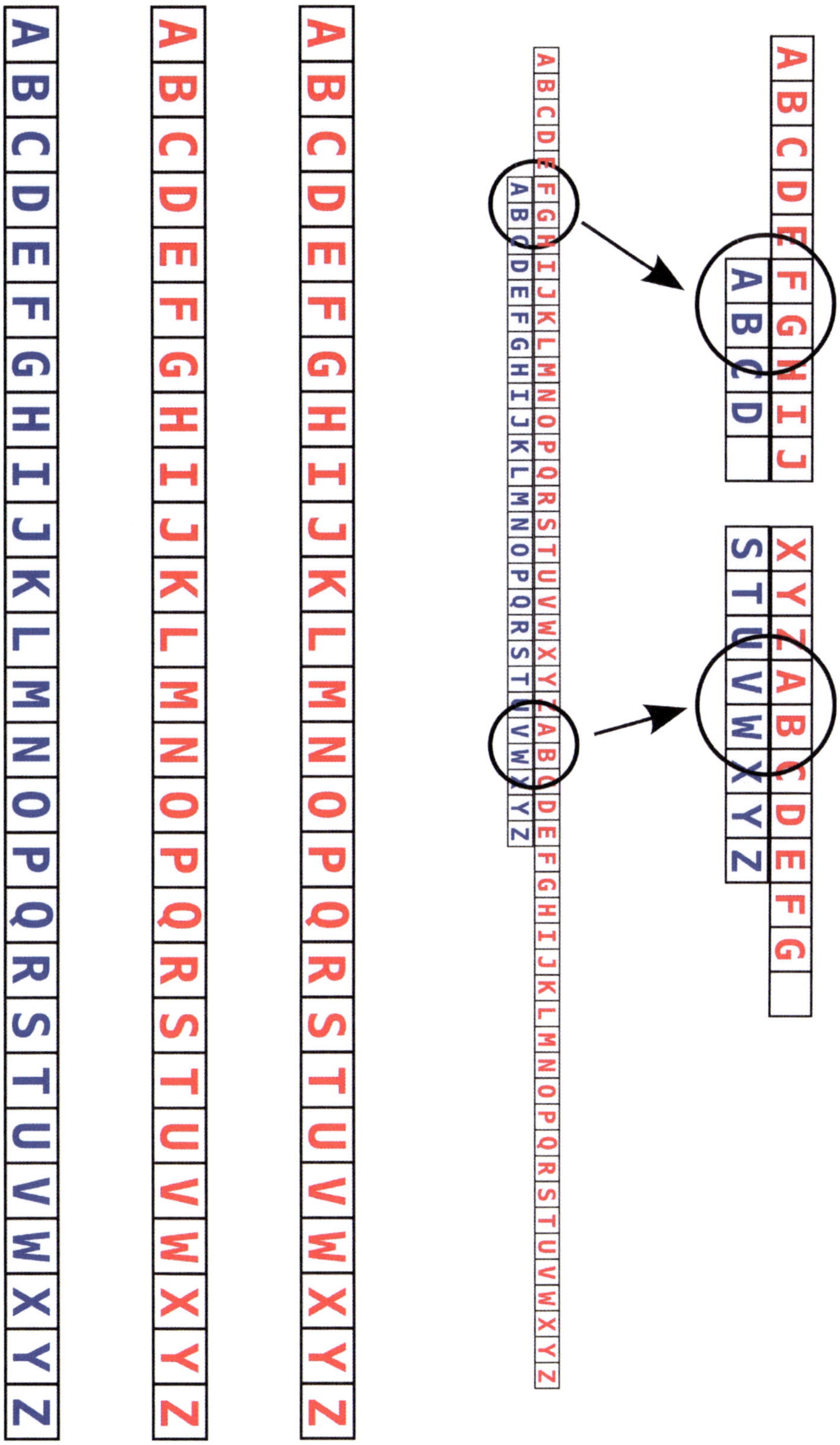
ABCDEFGHIJKLMNOPQRSTUVWXYZ
ABCDEFGHIJKLMNOPQRSTUVWXYZ
ABCDEFGHIJKLMNOPQRSTUVWXYZ
ABCDEFGHIJKLMNOPQRSTUVWXYZ
ABCDEFGHIJKLMNOPQRSTUVWXYZ
ABCDEFGHIJKLMNOPQRSTUVWXYZ
ABCDEFGHIJKLMNOPQRSTUVWXYZ

7 SYMBOL CIPHERS – PIGPEN CIPHER

There are some ciphers that are useful because you don't need to carry around special devices or charts to code or decode messages using them. One of these is called the Pig Pen cipher, since you remember how to use it by visualizing pigs and pig pens.

You basically draw two pig pens that resembles a tic-tac-toe board, and another two shaped like an **X**. You then populate one of each of these pens with pigs; well, actually just dots that represent pigs, as shown in the figure. You also write the alphabet letters in order, as shown.

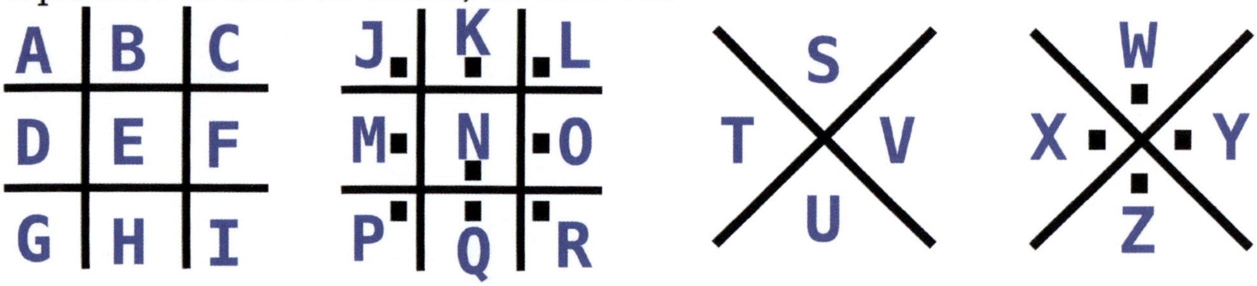

Your code is just the portion of a pig pen that each letter finds itself. You can write this out as a chart, as shown on the next page, or just find it in the figure.

Here are some examples of messages encoded using the Pig Pen cipher (Spaces are not coded).

PIG PEN CIPHER

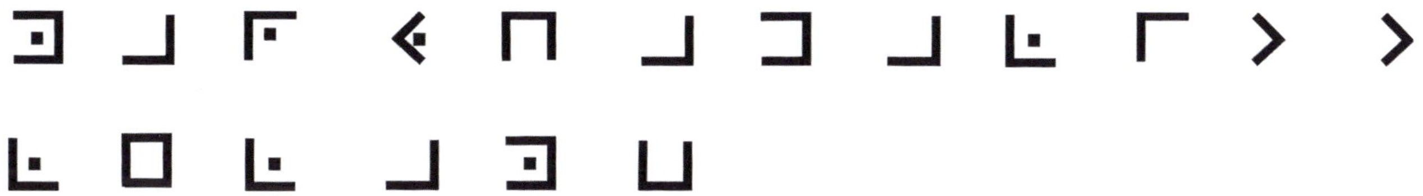

MARY HAD A LITTLE LAMB

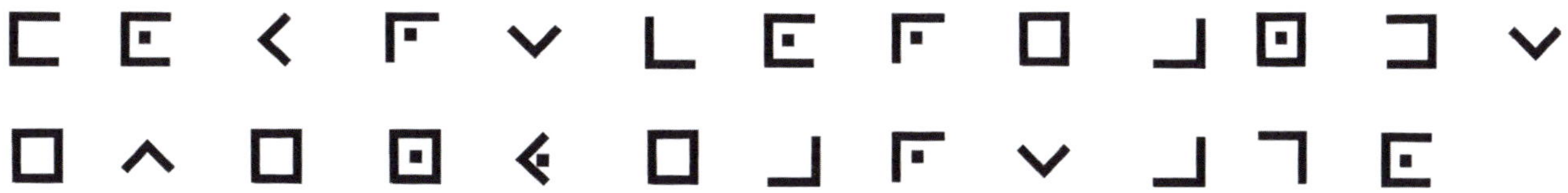

FOUR SCORE AND SEVEN YEARS AGO

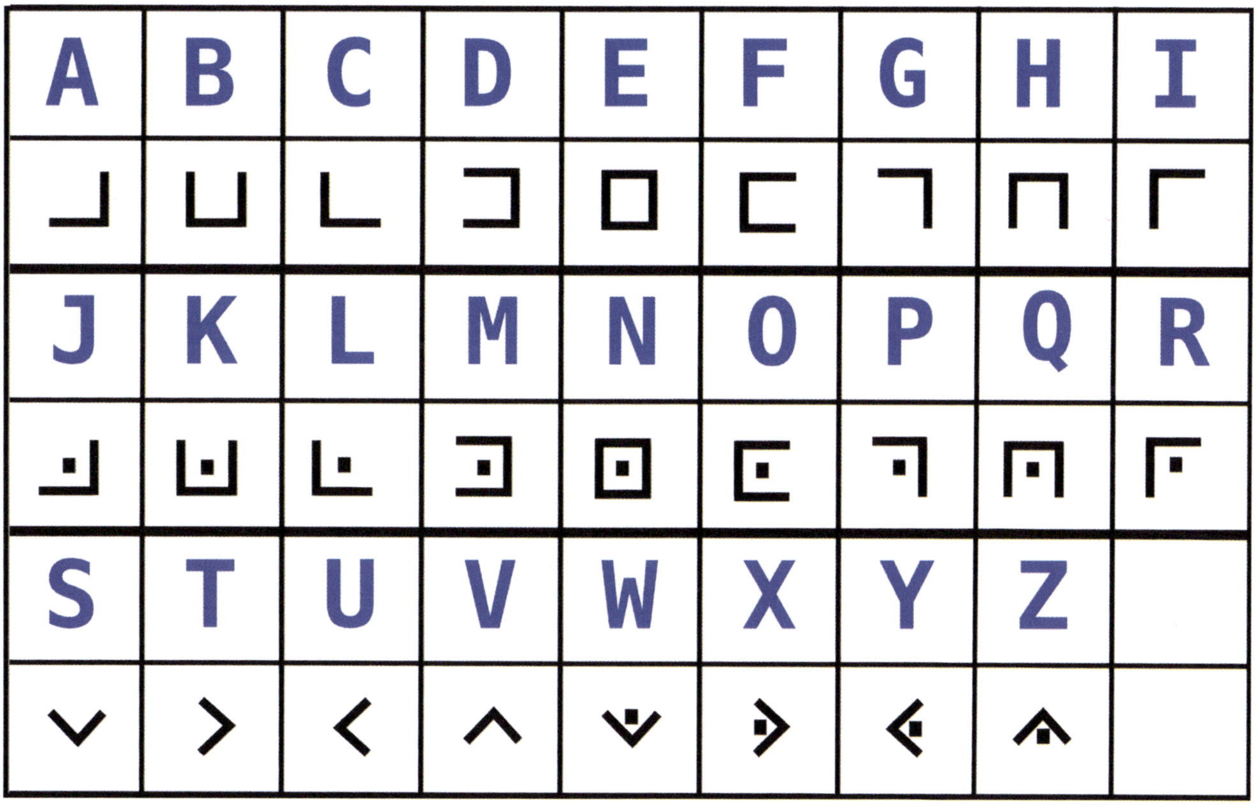

8 SYMBOL CIPHERS – FOREIGN ALPHABETS

A book written in a language you don't know is just like a code. People who understand the rules of the language are able to decipher the code.

Short of learning a new language, or inventing your own language, you can write messages that look like they're written in another language, but they aren't. In a process called transliteration, you substitute letters from the alphabet of another language for the letters in your message.

Transliteration is made easier by computer word processing programs that allow you to change the font of what you've written. You just highlight the text you want transliterated and select a foreign language font.

The charts that follow transliterate English characters into characters from four different languages; namely, Voynich, Greek, Hieroglyphic and Cyrillic fonts. Hieroglyphics are an ancient writing once used in Egypt, and Cyrillic is the font used in the Russian and Bulgarian languages.

The Voynich characters are unusual. They are found in the Voynich Manuscript, an ornately illustrated book in an unknown language. Cryptologists and linguists have studied the Voynich Manuscript for years, trying to decode it. They've found some properties that indicate that it's a real language, but they still don't have a translation. Here's what "Voynich Manuscript" looks like in the Voynich character set:

	Voynich	Greek
A		α
B	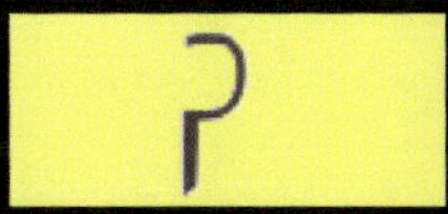	β
C		χ
D		δ
E		ε
F		φ
G		γ
H	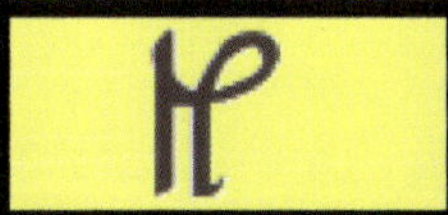	η
I		ι
J		ς
K		κ
L		λ
M		μ
N		ν
O		ο
P		π
Q		θ
R		ρ
S		σ
T		τ
U		υ
V		Δ
W		ω
X		ξ
Y		ψ
Z		ζ

	Hieroglyphics	Cyrillic
A		а
B		б
C		ъ
D		д
E		з
F		ф
G		г
H		ч
I		и
J		й
K		к
L		л
M		м
N		н
O		о
P		п
Q		я
R		р
S		с
T		т
U		у
V		в
W		ш
X		х
Y		ю
Z		ц

9 SYMBOL CIPHERS – MIRRORED/FLIPPED ALPHABET CODE

The famous artist, inventor, engineer and scientist, Leonardo da Vinci, is well known to technologists because of his superbly illustrated notebooks. One interesting thing about his notebooks is the way that he wrote his notes. In order to hide what he had written from prying eyes, he wrote everything backwards! His notes can be read by most people only when held up to a mirror.

As an example, hold this page up to a mirror. You'll see that what's below is a mirror image of the first paragraph of this chapter.

The famous artist, inventor, engineer and scientist, Leonardo da Vinci, is well known to technologists because of his superbly illustrated notebooks. One interesting thing about his notebooks is the way that he wrote his notes. In order to hide what he had written from prying eyes, he wrote everything backwards! His notes can be read by most people only when held up to a mirror.

Writing things backwards makes a message harder to read, but certainly not impossible. Another thing that could be done is to flip each letter individually. Such a flipped character image still can't be easily read when held up to a mirror, as this "Mary had a little lamb" text shows.

Mary had a little lamb

The chart on the next page illustrates a mirrored and flipped English alphabet.

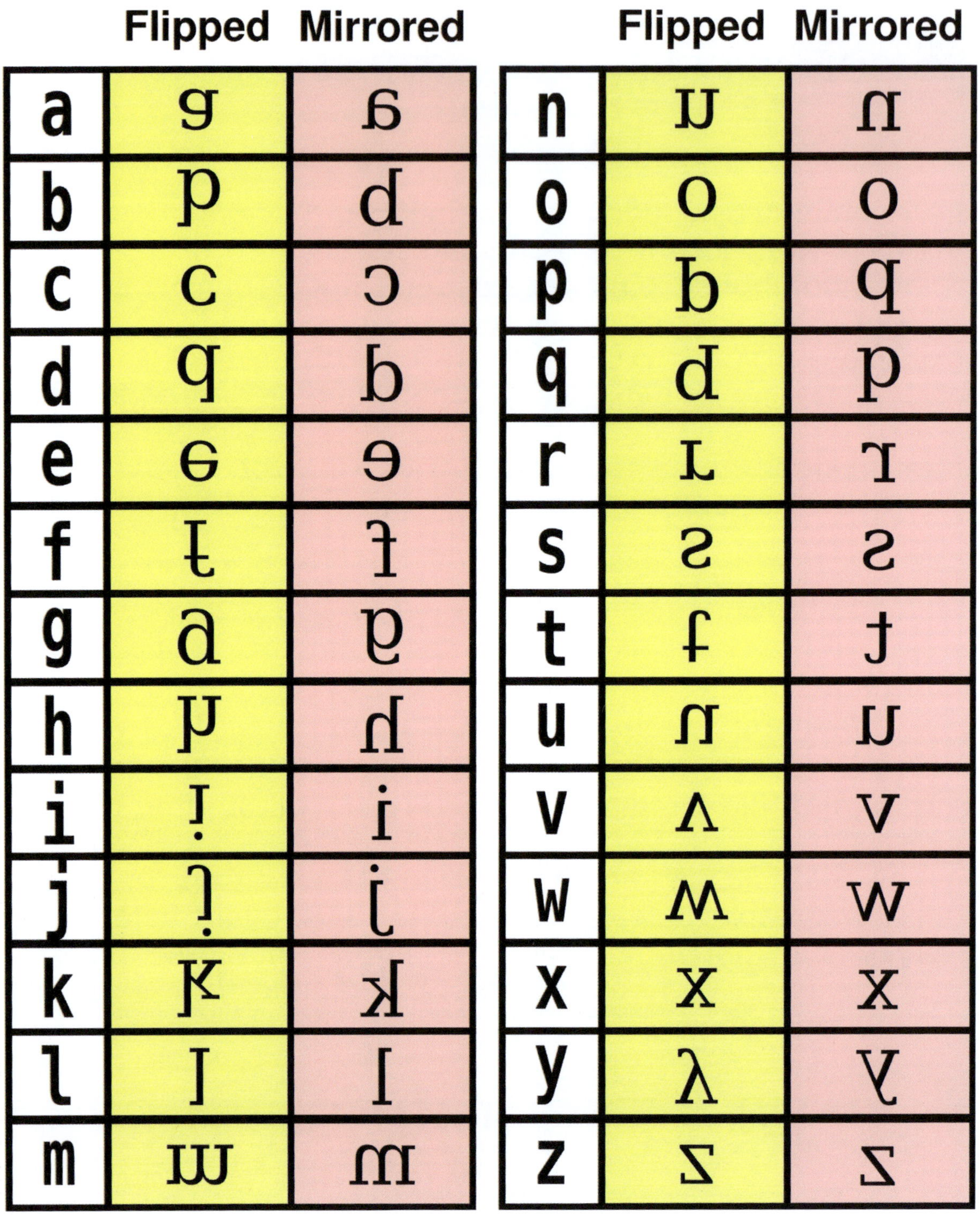
Flipped Mirrored
Flipped Mirrored
a
b
c
d
e
f
g
h
i
j
k
l
m
n
o
p
q
r
s
t
u
v
w
x
y
z

10 SYMBOL CIPHERS – MORSE CODE

Not all codes are designed to hide messages. Some codes are designed to make messages easier to send. One of these is the Morse Code, invented by the artist, Samuel F. B. Morse, and his colleague, Alfred Vail.

The Morse Code is designed to send messages in a simple electrical circuit. The circuit is so primitive that all it can do is turn things on and off. Letters in Morse Code are sent as combinations of a short "on" time (called a "short," "dit" or "dot") and a long "on" time (called a "long," "dah" or "dash).

Vail knew that some English letters were more common than others, a fact you'll read about in a later chapter. Vail knew that messages would be easier to send if the most used letters had the fewest number of dots and dashes.

Since the letter "e" is the most common letter, it's coded by a single dot. The next most common letter, "t," is coded by a single dash. The less common letter, "q," is coded by three dashes and one dot. Because of this scheme, the Morse Code allows a fast transmission of messages.

There's a chart of the Morse Code on the next page. The following is "Mary had a little lamb" in Morse Code.

– –   · –   · – ·   – · – –   · · · ·   · –   – · ·   · –   · – · ·

· ·   –   –   · – · ·   ·   · – · ·   · –   – –   – · · ·

A • ▬
B ▬ • • •
C ▬ • ▬ •
D ▬ • •
E •
F • • ▬ •
G ▬ ▬ •
H • • • •
I • •
J • ▬ ▬ ▬
K ▬ • ▬
L • ▬ • •
M ▬ ▬
N ▬ •
O ▬ ▬ ▬
P • ▬ ▬ •
Q ▬ ▬ • ▬
R • ▬ •
S • • •
T ▬
U • • ▬
V • • • ▬
W • ▬ ▬
X ▬ • • ▬
Y ▬ • ▬ ▬
Z ▬ ▬ • •

1 • ▬ ▬ ▬ ▬
2 • • ▬ ▬ ▬
3 • • • ▬ ▬
4 • • • • ▬
5 • • • • •
6 ▬ • • • •
7 ▬ ▬ • • •
8 ▬ ▬ ▬ • •
9 ▬ ▬ ▬ ▬ •
0 ▬ ▬ ▬ ▬ ▬

11 SYMBOL CIPHERS – BRAILLE

Braille was an important communications medium for the blind before optical character recognition and text-to-speech systems became inexpensive and ubiquitous. The Braille code consists of small patches of raised dots. A skilled Braille reader can rapidly scan his fingers along lines of these patches to read text as easily as a sighted person reads a printed book..

When using Braille as a code, it isn't necessary to create the raised dots. Printed dots will work as well. The table on the next page shows the alphabet as Braille code.

"Mary had a little lamb" looks like this in Braille code.

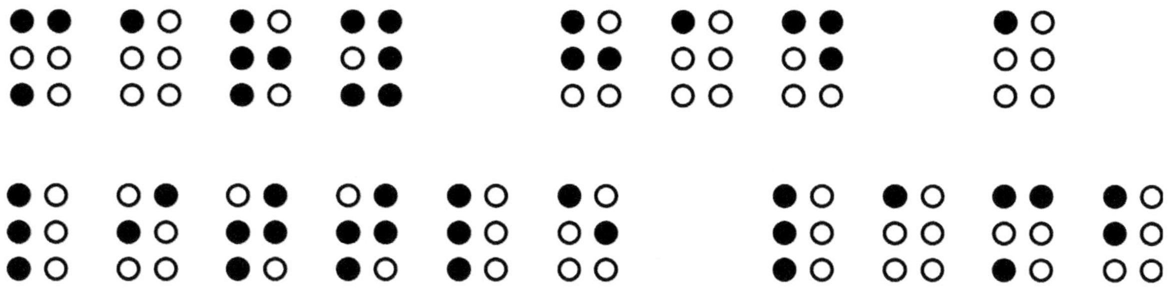

a	⠁
b	⠃
c	⠉
d	⠙
e	⠑
f	⠋
g	⠛
h	⠓
i	⠊
j	⠚
k	⠅
l	⠇
m	⠍
n	⠝
o	⠕
p	⠏
q	⠟
r	⠗
s	⠎
t	⠞
u	⠥
v	⠧
w	⠺
x	⠭
y	⠽
z	⠵

12 SYMBOL CIPHERS – SEMAPHORES

How did people communicate with each other over distances in the days before telegraphs, telephones and radio? For long distances, messengers were the only option, and that was one reason why people became interested in cryptography. Letters could be opened and read at many places along a messenger's journey.

Other methods were developed for closer distances, such as between troops on a battlefield, ships at sea, and ship to shore. For example, lights or flashes of sunlight from mirrors can be used to transmit the Morse Code described in a previous chapter. Smoke from fires can also be used to send primitive signals long distances. Drums and horns were also used as message systems by primitive people.

One very simple and effective communication device is the flag. Waving a single flag might get someone's attention, but you need two flags and a little finesse to send real messages. The flag semaphore system makes use of two flags, held by a person in his right and left hands, to send characters through different hand positions.

The chart on the next page shows the hand positions for the alphabet characters. You can also write these out to code a secret message. "Mary had a little lamb" looks like this in semaphore code.

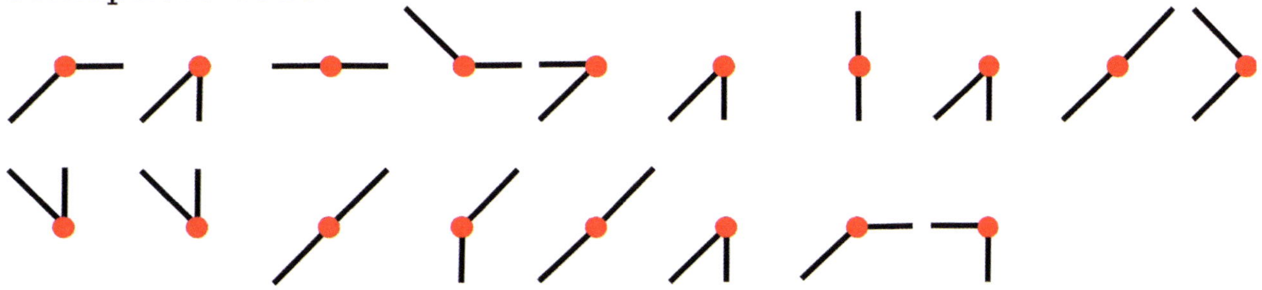

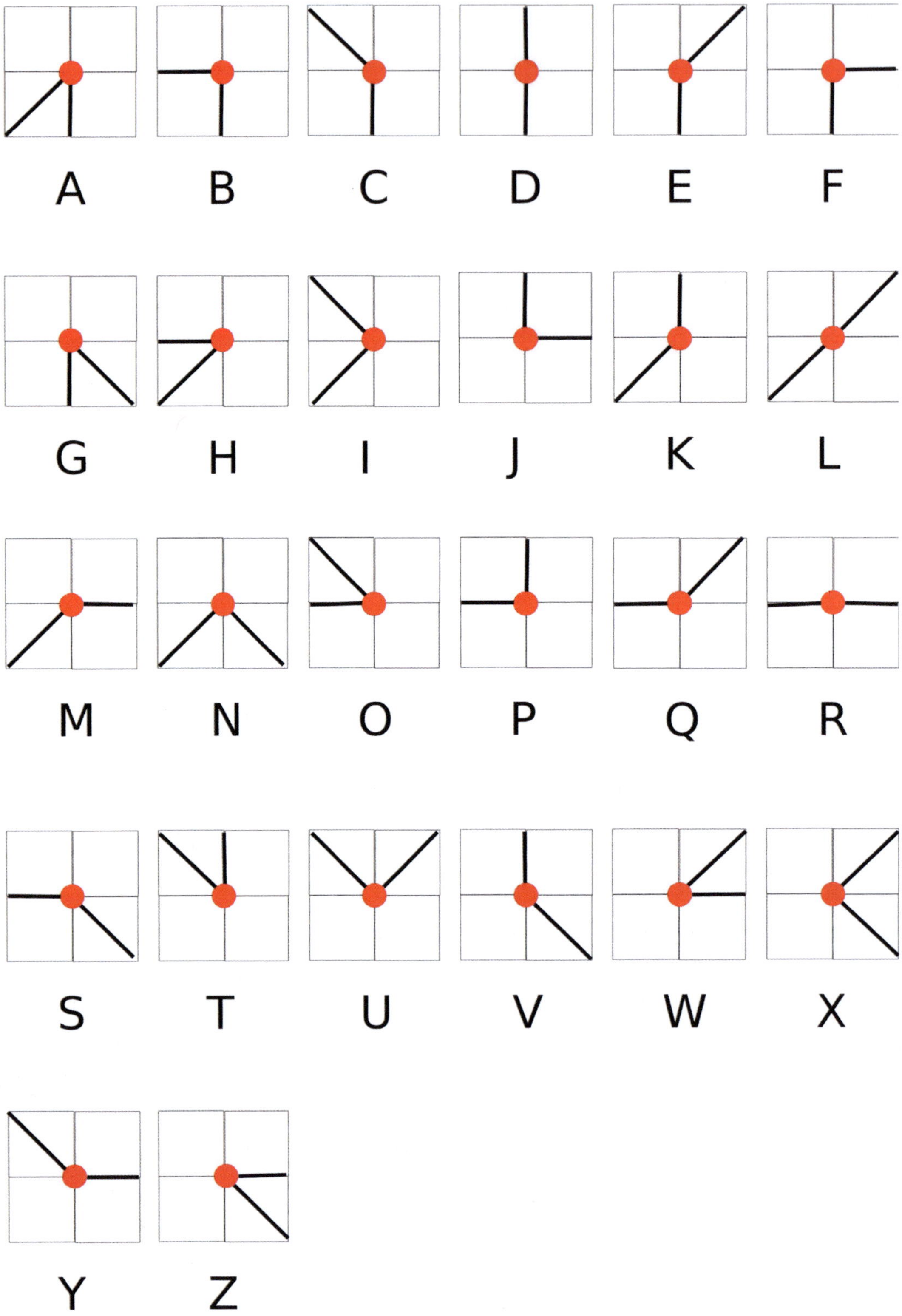
A
B
C
D
E
F
G
H
I
J
K
L
M
N
O
P
Q
R
S
T
U
V
W
X
Y
Z

13 SYMBOL CIPHERS – FOURTEEN SEGMENT CODE

In the early days of electronic devices, before computer display screens became common, messages were often displayed on indicators for which the alphabet was rendered using fourteen line segments. You can see how this was done in the figure on the next page.

You can use the letter segments as a code for writing secret messages. Unlike the Morse Code, which tries to use the shortest length code to send messages, the fourteen segment code uses a lot of numbers to code each character, as shown in the following chart.

A	1	2	3	5	6	7	8
B	1	2	3	4	8	10	13
C	1	4	5	6			
D	1	2	3	4	10	13	
E	1	4	5	6	7	8	
F	1	5	6	7	8		
G	1	3	4	5	6	8	
H	2	3	5	6	7	8	
I	1	4	10	13			
J	2	3	4	5			
K	5	6	7	11	14		
L	4	5	6				
M	2	3	5	6	9	11	

N	2	3	5	6	9	14	
O	1	2	3	4	5	6	
P	1	2	5	6	7	8	
Q	1	2	3	4	5	6	14
R	1	2	5	6	7	8	14
S	1	3	4	6	7	8	
T	1	10	13				
U	2	3	4	5	6		
V	5	6	11	12			
W	2	3	5	6	12	14	
X	9	11	12	14			
Y	9	11	13				
Z	1	4	11	12			

Since the highest numbered segment is fourteen, you can use any number greater than that to act as a space between character codes.

"Mary" looks like this in fourteen segment code.

2,3,5,6,9,11,20,1,2,3,5,6,7,8,20,1,2,5,6,7,8,14,15,9,11,13,18

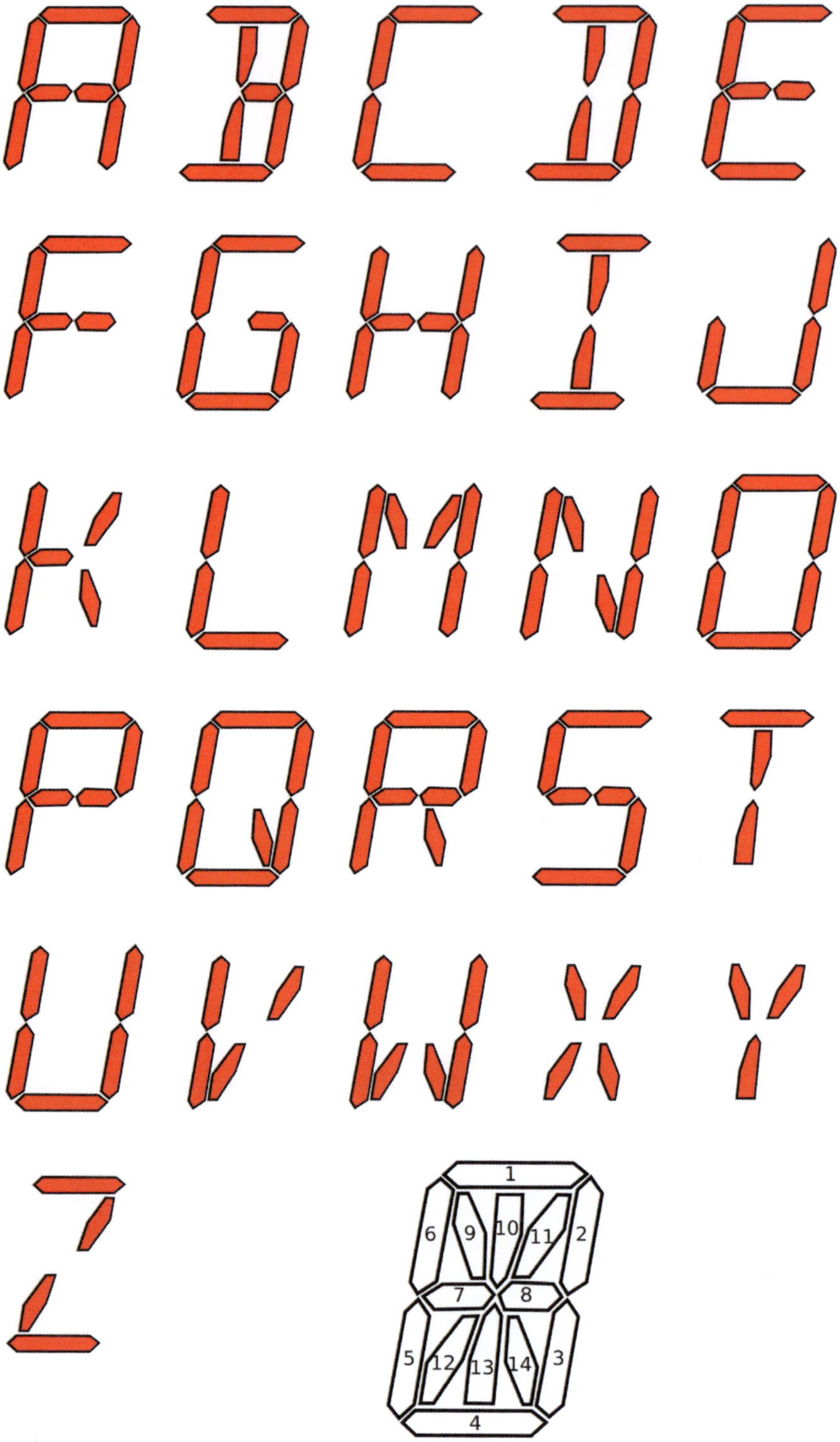
1
6
9
10
11
2
7
8
5
12
13
14
3
4

14 SYMBOL CIPHERS – SAY IT WITH FLOWERS

What can be more innocent than a drawing of a flower? The petals on a flower can hide your secret message. The semaphore code in a previous chapter can be represented using a flower.

Since the semaphore code for a character just has two flags, this would translate to a flower with just two petals, which wouldn't look like a real flower. The best approach, as shown in the table on the next page, is to have the semaphore code done as the absence of a petal.

Such a code is still readily readable, as the following "Mary had a little lamb" example shows.

"Mary had a little lamb" in flower code.

A B C D E F
G H I J K L
M N O P Q R
S T U V W X
Y Z

15 SYMBOL CIPHERS – SAY IT WITH MUSIC

Just as other symbols can be used to write secret messages, you can do the same with musical notes. In this case, the written music (called a "score") can be sent as the message; or, if the sender and receiver are skilled musicians, the actual music as sound can be sent.

First, you need to remember how musical notes are written. The following example shows the seven notes of the most common musical scale (C-major).

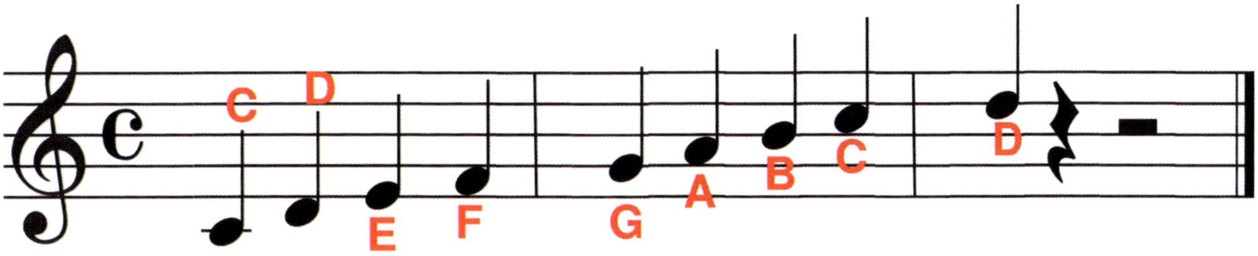

Notes are not always played alone. They are often played in pairs, or as three or more notes at a time. These are known as chords. One common chord is illustrated below. Aside from the notes, there are other symbols called rests. These are needed for an accurate representation of the music.

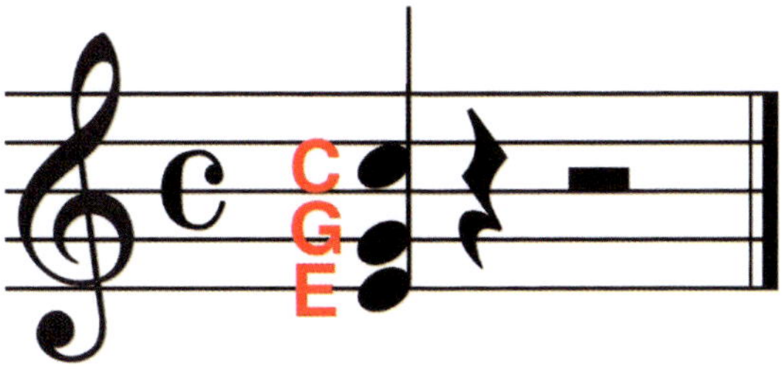

Some notes sound nice in chords, and some don't. If you were going to make a code using musical notes, you would try to make the notes sound like music, and not like a cat walking on a piano keyboard.

An obvious code choice for the letters A-G would be the musical notes A-G. After that, it gets a little harder. Here's a selection of notes and chords that sound musical.

Notes:	A B C D E F G
Thirds:	C-E E-G G-B B-D F-A A-C
Fifths:	C-G G-D D-A A-E B-F# F#-C# C#-G# G#-D# D#-A# A#-F
Triads:	C-E-G E-G-B G-B-D

Coding is easier, since there's more than one type of each musical note based on how long the note should sound. Notes can be quarter notes, half notes, or other types.

The music on the next page shows one way to code letters as music. Here's "Mary had a little lamb" in musical code.

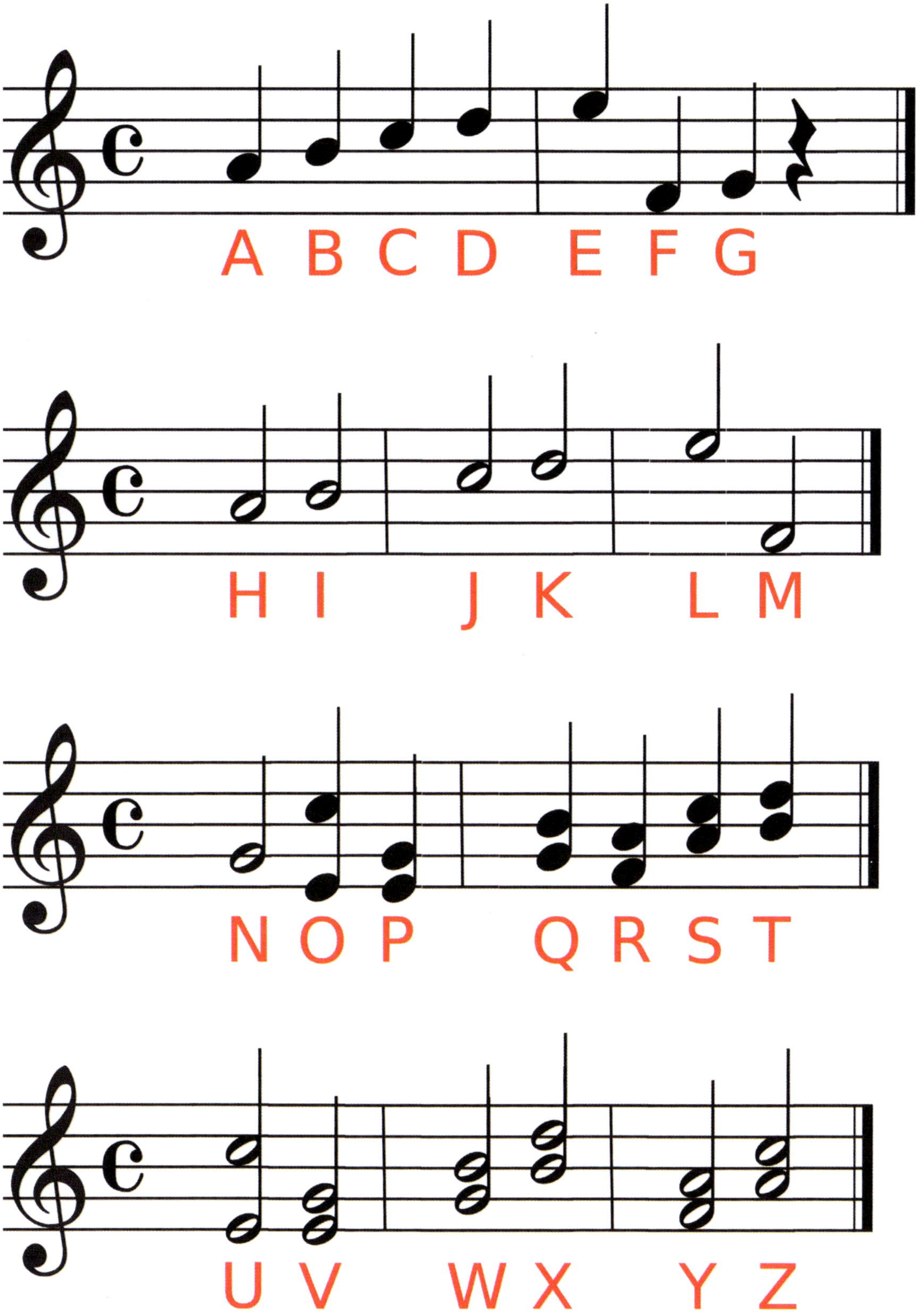
A B C D E F G
H I J K L M
N O P Q R S T
U V W X Y Z

16 SYMBOL CIPHERS – SAY IT WITH COLORS

You don't need symbols to write a coded message. You can use colors. Color codes are actually common for many things. Electrical components were once labeled using colors.

The reason for this was simple. It's much easier just to paint dots or stripes of color than numbers. Some of these components were small, so a small number printed on them would be hard to print and read, and these numbers could be easily damaged.

The simplest color code was for the numbers zero through nine, as shown below.

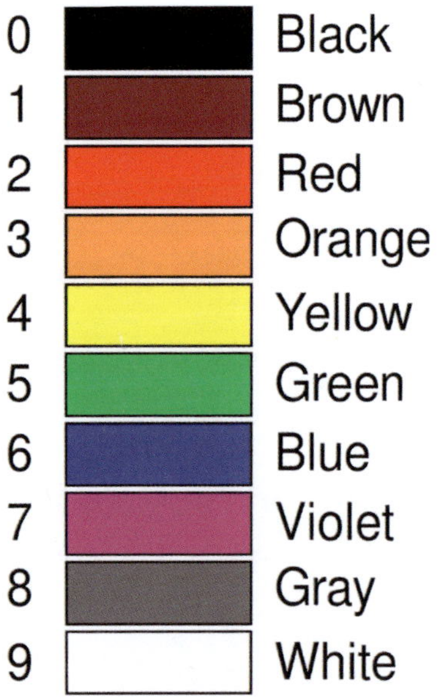

With a careful choice of colors, we can distinguish all twenty six letters in the English alphabet, as shown in the chart on the next page.

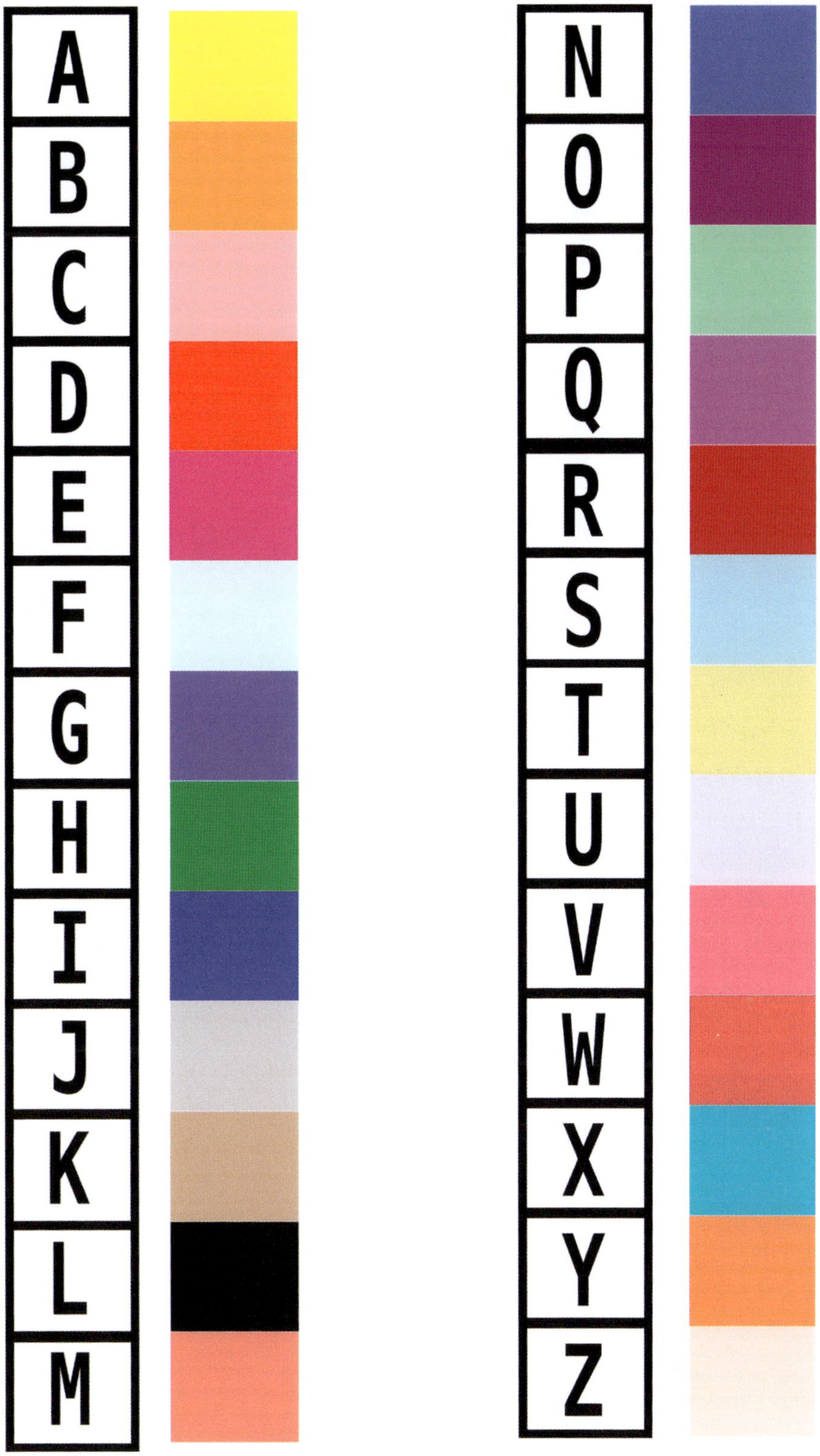
A
B
C
D
E
F
G
H
I
J
K
L
M
N
O
P
Q
R
S
T
U
V
W
X
Y
Z

17 SYMBOL CIPHERS – SAY IT WITH PIGGIES

Small changes to the shape, color or composition of an object might not be perceptible to a casual viewer. However, difference in an image can convey a code.

As an example of this, look at the following drawings of a cartoon pig.

If you look closely, you'll see a lot of difference between the two pigs. The eyebrows are different, and one has a drooping ear, a spot, and an extra curl in his tail.

As the chart on the next page shows, you can use such changes to make twenty six different pigs to give you an alphabet for codes.

Here's what "Mary had a little lamb" looks like in piggie code.

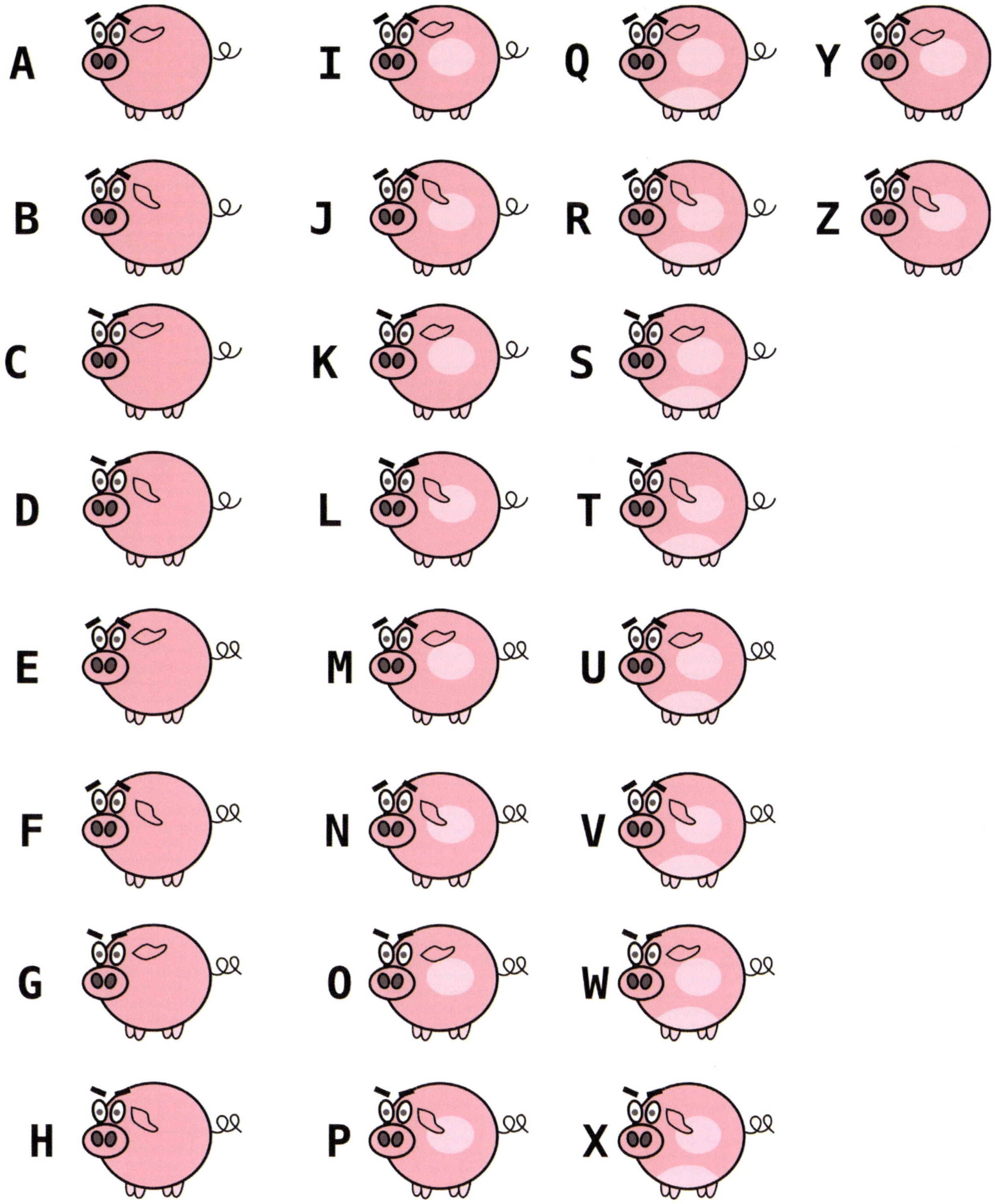
A
I
Q
Y
B
J
R
Z
C
K
S
D
L
T
E
M
U
F
N
V
G
O
W
H
P
X

18 SYMBOL CIPHERS – COLOR CODED IMAGES

In a previous chapter, we saw how a set of twenty six colors distinguishable by eye can be used to make coded messages. One advantage of computers is that they can do some things better than people. One of these things is the ability to distinguish colors in images.

Images we see on most computers can contain an amazing 16,777,216 colors. This gives people a phenomenal ability to hide messages in normal looking or random images.

Using a computer, it's easy to make an image that codes a message only in shades of red.

The following image demonstrates how the phrase, "Mary had a little lamb whose fleece was white as snow, and everywhere that Mary went, the lamb was sure to go," can be coded in shades of red.

19 SYMBOL CIPHERS – COMPUTER CODES

Computers are able to do things such as displaying text on monitors and doing word processing because the letters are stored as a code.

These codes are not new with computers. The first such code for electrical communication was invented in 1870. Text messages have been transmitted on wires and by radio using such codes for more than a hundred years.

If we were to limit ourselves to a twenty six character alphabet, we could used the numbers 1-26, or 0-25 as a code. To make computers as versatile as possible, the most common character code of early computers used the numbers 0-127 to include everything from upper and lower case letters, numbers, some symbols, and even bell ringing. This was the American Standard Code for Information Interchange (ASCII).

The chart on the next page lists the ASCII codes for the upper case letters in both decimal and hexadecimal (base-16) form. "Mary had a little lamb, whose fleece was white as snow" looks like this in computer code.

```
77 65 82 89 72 65 68 65 76 73 84 84 76 69 76 65 77 66
 M  A  R  Y  H  A  D  A  L  I  T  T  L  E  L  A  M  B

87 72 79 83 69 70 76 69 69 67 69 87 65 83 87 72 73 84 69
 W  H  O  S  E  F  L  E  E  C  E  W  A  S  W  H  I  T  E

65 83 83 78 79 87
 A  S  S  N  O  W
```

	Decimal	Hex
A	65	41
B	66	42
C	67	43
D	68	44
E	69	45
F	70	46
G	71	47
H	72	48
I	73	49
J	74	4A
K	75	4B
L	76	4C
M	77	4D

	Decimal	Hex
N	78	4E
O	79	4F
P	80	50
Q	81	51
R	82	52
S	83	53
T	84	54
U	85	55
V	86	56
W	87	57
X	88	58
Y	89	59
Z	90	5A

20 SIMPLE CIPHERS - PIG LATIN

The simple word game, Pig Latin, actually qualifies as a permutation cipher. That's because its simplest operation, coding a word that begins with a consonant, involves moving the first letter of the word to the end. There's an added operation, called obfuscation, in which the letters "ay" are added to the end of this new word.

Words starting with a vowel are treated differently, as the operations for our "Mary had a little lamb" example show. For those words, you just add "way" to the end of the word.

Mary	→	Ary-may
had	→	ad-hay
a	→	a-way
little	→	ittle-lay
lamb	→	amb-lay

Doing a Pig Latin operation on "Mary had a little lamb" gives us this.

Ary-may ad-hay a-way ittle-lay amb-lay

21 SIMPLE CIPHERS - CAESAR CIPHER

We've seen the operation of the Caesar cipher in the chapter about Caesar rods. This cipher is named after the Roman Emperor, Julius Caesar, who started his career as a general. Like all generals, he needed to send orders to his troops. To keep these orders secret, he used a cipher he didn't actually invent, but it's named after him since he was famous.

The Caesar cipher works by substituting other alphabet characters for the characters in your message, but it does this in a very simple way. To encode a message using the Caesar cipher, you choose a number from 1 to 25. This number is used to select the substitute letters for letters in your message.

The example on the next page shows how to encode the word, "DISPLAY," using a Caesar cipher based on the number four. For each letter in your message, you count down the alphabet four places to find the coded letter.

D → H
I → M
S → W
P → T
L → P
A → E
Y → C

The tricky part is at Y, when we count off the end of the alphabet. In that case, we just start counting again at the start.

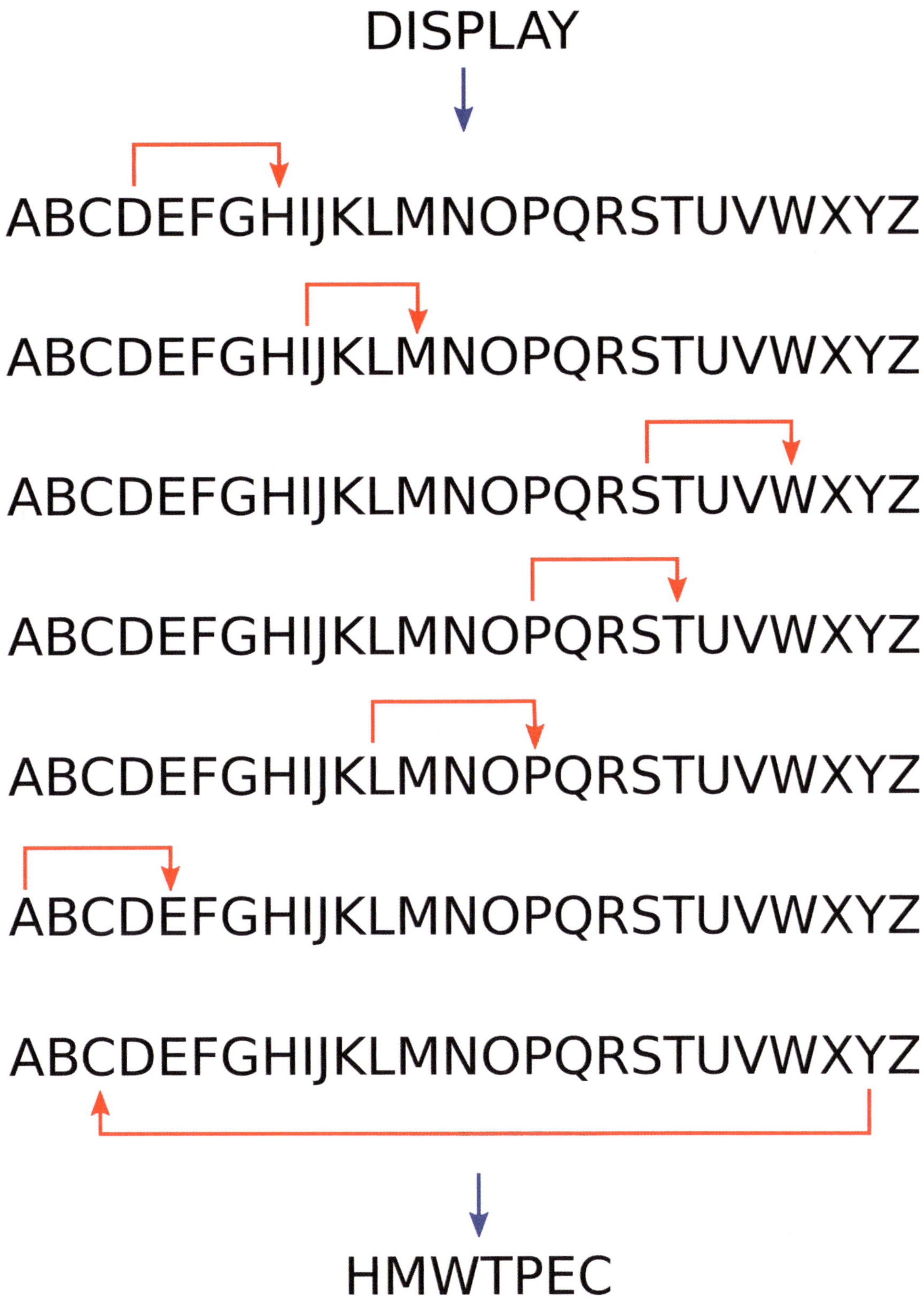
DISPLAY
ABCDEFGHIJKLMNOPQRSTUVWXYZ
ABCDEFGHIJKLMNOPQRSTUVWXYZ
ABCDEFGHIJKLMNOPQRSTUVWXYZ
ABCDEFGHIJKLMNOPQRSTUVWXYZ
ABCDEFGHIJKLMNOPQRSTUVWXYZ
ABCDEFGHIJKLMNOPQRSTUVWXYZ
ABCDEFGHIJKLMNOPQRSTUVWXYZ
HMWTPEC

22 SIMPLE CIPHERS - ONE-TIME PAD

A one-time pad is an essentially unbreakable cipher; provided, however, that the "pad" is never revealed.

In the days before computers, the pad was a page of random characters, usually bound into a book. Each page of the book was marked with a code word, or a number, so that the sender and receiver knew what page was used to encode the message. The pad was know as a "one-time pad," since the page was to be destroyed after encoding a message.

In the computer age, the pad characters can be non-printing characters, such as the "escape key," giving a wider range of possible codes. There are many ways to implement the encoding of a one-time pad cipher. In our example we'll use a scheme based on the Caesar cipher in the last chapter.

The first step is generation of the pad. For this you can write the alphabet on pieces of paper, shake them up in a bag, and pick random slips one at a time. After writing down the letter, you put the piece of paper back into the bag, shake again, and continue for as many characters your message has. Here's an example of a one-time pad generated by a computer:

RSQPBELOMGXEWHQTMLZNCVDUXPKG

After you have the pad, you'll need to give a copy to whomever will receive your coded message. Since you usually don't know how long your messages will be, it's good to make a pad with more than enough letters for any message.

For our message, we'll use the simple sentence, "This is an example of a one-time pad." Of course, as we've seen in previous chapters, messages are easier to encode if they just consist of upper-case characters, without punctuation. You can still understand the sentence when it's written that way:

THISISANEXAMPLEOFAONETIMEPAD

The message is coded using the Caesar cipher in the following way. We see that the first letter of our pad, R, is the eighteenth letter of the alphabet. This means we should encode the first letter of our message using a Caesar cipher in which the letters are shifted eighteen places, so our T becomes a K.

Moving on to the second character, our pad has the letter S, so we need to do a Casar cipher with a shift of nineteen. The H in our message becomes a Z. We continue through the entire message that way, with the following result:

KZYHJWLBQDXQLSUHRLNAGOLGBEKJ

All this is a little easier if you use the chart below, noting the pad character in red, the message character in blue, and the resulting coded character in black.

Decoding is done by performing the operations again on the coded message using the one-time pad. The first character in the coded message is K, and the pad character is R. Scanning across the row of the red R in the chart, we find K, and then look to the top of the column to see that the first letter, in blue, is T.

	A	B	C	D	E	F	G	H	I	J	K	L	M	N	O	P	Q	R	S	T	U	V	W	X	Y	Z
A	A	B	C	D	E	F	G	H	I	J	K	L	M	N	O	P	Q	R	S	T	U	V	W	X	Y	Z
B	B	C	D	E	F	G	H	I	J	K	L	M	N	O	P	Q	R	S	T	U	V	W	X	Y	Z	A
C	C	D	E	F	G	H	I	J	K	L	M	N	O	P	Q	R	S	T	U	V	W	X	Y	Z	A	B
D	D	E	F	G	H	I	J	K	L	M	N	O	P	Q	R	S	T	U	V	W	X	Y	Z	A	B	C
E	E	F	G	H	I	J	K	L	M	N	O	P	Q	R	S	T	U	V	W	X	Y	Z	A	B	C	D
F	F	G	H	I	J	K	L	M	N	O	P	Q	R	S	T	U	V	W	X	Y	Z	A	B	C	D	E
G	G	H	I	J	K	L	M	N	O	P	Q	R	S	T	U	V	W	X	Y	Z	A	B	C	D	E	F
H	H	I	J	K	L	M	N	O	P	Q	R	S	T	U	V	W	X	Y	Z	A	B	C	D	E	F	G
I	I	J	K	L	M	N	O	P	Q	R	S	T	U	V	W	X	Y	Z	A	B	C	D	E	F	G	H
J	J	K	L	M	N	O	P	Q	R	S	T	U	V	W	X	Y	Z	A	B	C	D	E	F	G	H	I
K	K	L	M	N	O	P	Q	R	S	T	U	V	W	X	Y	Z	A	B	C	D	E	F	G	H	I	J
L	L	M	N	O	P	Q	R	S	T	U	V	W	X	Y	Z	A	B	C	D	E	F	G	H	I	J	K
M	M	N	O	P	Q	R	S	T	U	V	W	X	Y	Z	A	B	C	D	E	F	G	H	I	J	K	L
N	N	O	P	Q	R	S	T	U	V	W	X	Y	Z	A	B	C	D	E	F	G	H	I	J	K	L	M
O	O	P	Q	R	S	T	U	V	W	X	Y	Z	A	B	C	D	E	F	G	H	I	J	K	L	M	N
P	P	Q	R	S	T	U	V	W	X	Y	Z	A	B	C	D	E	F	G	H	I	J	K	L	M	N	O
Q	Q	R	S	T	U	V	W	X	Y	Z	A	B	C	D	E	F	G	H	I	J	K	L	M	N	O	P
R	R	S	T	U	V	W	X	Y	Z	A	B	C	D	E	F	G	H	I	J	K	L	M	N	O	P	Q
S	S	T	U	V	W	X	Y	Z	A	B	C	D	E	F	G	H	I	J	K	L	M	N	O	P	Q	R
T	T	U	V	W	X	Y	Z	A	B	C	D	E	F	G	H	I	J	K	L	M	N	O	P	Q	R	S
U	U	V	W	X	Y	Z	A	B	C	D	E	F	G	H	I	J	K	L	M	N	O	P	Q	R	S	T
V	V	W	X	Y	Z	A	B	C	D	E	F	G	H	I	J	K	L	M	N	O	P	Q	R	S	T	U
W	W	X	Y	Z	A	B	C	D	E	F	G	H	I	J	K	L	M	N	O	P	Q	R	S	T	U	V
X	X	Y	Z	A	B	C	D	E	F	G	H	I	J	K	L	M	N	O	P	Q	R	S	T	U	V	W
Y	Y	Z	A	B	C	D	E	F	G	H	I	J	K	L	M	N	O	P	Q	R	S	T	U	V	W	X
Z	Z	A	B	C	D	E	F	G	H	I	J	K	L	M	N	O	P	Q	R	S	T	U	V	W	X	Y

23 SIMPLE CIPHERS - BOOK CIPHER

If you and your friend have access to the same book, which is very easy when the book is old enough to be on the Internet, you can send very secure coded messages using a book cipher.

The idea behind a book cipher is that the location of letters in the book can be used instead of the letters themselves. If the first letter of your message is a "T," you choose some random "T" in the book and note its position.

If it's a printed book, the position code can be the page number, combined with the sentence number, word number, and letter number in the word. If it's a digital book on the Internet, another method is needed.

The text on the next page is from the book, The Legend of Sleepy Hollow, by Washington Irving. One letter of the text was modified, so that the text includes all alphabet characters ("j" was substituted for "g" in the word, "magic").

As you can see, all spaces and punctuation have been removed, and all letters changed to upper case. The letters have been arranged in rows of fifty characters each. To code a letter, you write its row number and column number.

Our example phrase, "Mary had a little lamb," would be coded as follows.

16-36 9-5 13-44 12-20 2-42 15-9 5-45 14-28 17-36 9-21
8-25 14-47 11-6 18-26 4-32 3-3 19-2 2-5

```
           10        20        30        40        50
1  THEBROOKWASSEARCHEDBUTTHEBODYOFTHESCHOOLMASTERWASN
2  OTTOBEDISCOVEREDHANSVANRIPPERASEXECUTOROFHISESTATE
3  EXAMINEDTHEBUNDLEWHICHCONTAINEDALLHISWORLDLYEFFECT
4  STHEYCONSISTEDOFTWOSHIRTSANDAHALFTWOSTOCKSFORTHENE
5  CKAPAIRORTWOOFWORSTEDSTOCKINGSANOLDPAIROFCORDUROYS
6  MALLCLOTHESARUSTYRAZORABOOKOFPSALMTUNESFULLOFDOGSE
7  ARSANDABROKENPITCHPIPEASTOTHEBOOKSANDFURNITUREOFTH
8  ESCHOOLHOUSETHEYBELONGEDTOTHECOMMUNITYEXCEPTINGCOT
9  TONMATHERSHISTORYOFWITCHCRAFTANEWENGLANDALMANACAND
10 ABOOKOFDREAMSANDFORTUNETELLINGINWHICHLASTWASASHEET
11 OFFOOLSCAPMUCHSCRIBBLEDANDBLOTTEDINSEVERALFRUITLES
12 SATTEMPTSTOMAKEACOPYOFVERSESINHONOROFTHEHEIRESSOFV
13 ANTASSELTHESEMAJICBOOKSANDTHEPOETICSCRAWLWEREFORTH
14 WITHCONSIGNEDTOTHEFLAMESBYHANSVANRIPPERWHOFROMTHAT
15 TIMEFORWARDDETERMINEDTOSENDHISCHILDRENNOMORETOSCHO
16 OLOBSERVINGTHATHENEVERKNEWANYGOODCOMEOFTHISSAMEREA
17 DINGANDWRITINGWHATEVERMONEYTHESCHOOLMASTERPOSSESSE
18 DANDHEHADRECEIVEDHISQUARTERSPAYBUTADAYORTWOBEFOREH
19 EMUSTHAVEHADABOUTHISPERSONATTHETIMEOFHISDISAPPEARA
20 NCE
```

24 SIMPLE CIPHERS - BOOK STENCIL CIPHER

The book stencil cipher is a lot like the book cipher in the previous chapter. If your message is short, you might be able to find all the letters of your message, in order, on a book page.

To encode your message, you make a stencil. A stencil is a sheet of paper with cut, or punched, holes. The idea is to create a hole in the stencil over the letter you want in the book. You punch holes in the stencil for all your message letters.

When the stencil is placed over the page in your book, or the same book owned by the person who receives your stencil, the message is revealed. People who don't know the book or page used to make the stencil can't decipher the message.

Longer book stencil messages are possible using computers. A computer can mix your message letters with random letters and print a book page. It can also make a companion stencil for the page. Anyone who has both the computer-printed page and the stencil can read the message.

Usually the computer-printed page and stencil are sent using different communication channels. For example, the page can be sent as an attachment to an email message, and the stencil can be mailed or hand-delivered.

The example on the next page shows the message, “this is a book stencil cipher,” coded as a book stencil cipher.

wy9wbd1h61u01ao26z7ultoy5safcg
9fy5qc77j92jad7ck6jehd2piu0y7h
gwmmmztui3w1da8puf986nbwi6x5de
ubghxfbsxepyrzshtmq2sddpjjmox1
icx9ack3ays8qrg712xnvb1ku17jj9
bmykwb5013suuka15dytsaut4g1nza
bwnvty0tk4mfy6q0b3z5ca06dg71gi
redn8bo1xsncru2r66rbj0bpesxg9g
0ukn3wb5t4s6wnootff4zggou9mai2
ix160yh5rmyxim67fmmko20axax7sz
bxqhyoz4g13e4p1b47yps0q3xk05s0
4x11w910q744btpnsx1nqmdtjdtp6u
as5wofip5pta1m0p1xa8nxthhzb655
gxvdmavbg15qmrqcdom1ck6t1548e9
r7u4t4051lab822kfr703r7xtd11pn
giulczmik7swguz0aq8btzirscsul4
2nooznah8ia4y53zft7qi79e057uqb
sydg10rg89qd8vg7aae1xlaclfjg6v
zticsi78rhy5pt1w4wb5uwf6ivqz1m
0fmg75ogx5e2u7oqci1heyx1hscgty
te7amcyp94gcyo0p0kqizpfk77wo0u
u9g5vt4x38df61hv20mk0czkj6h87c
2ly7grbdvjraqtqmvr9gr1fc5meeu7
wsu39mfnhyerp7ib3arj26zfcm9qgh
c9ggj2v7jt1bjtuux4fg61gly2voic

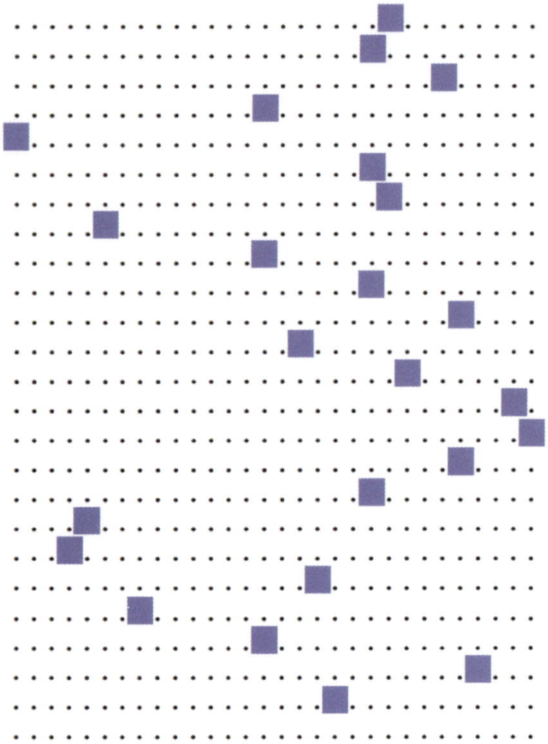

wy9wbd1h61u01ao26z7ultoy5safcg
9fy5qc77j92jad7ck6jehd2piu0y7h
gwmmmztui3w1da8puf986nbwi6x5de
ubghxfbsxepyrzshtmq2sddpjjmox1
icx9ack3ays8qrg712xnvb1ku17jj9
bmykwb5013suuka15dytsaut4g1nza
bwnvty0tk4mfy6q0b3z5ca06dg71gi
redn8bo1xsncru2r66rbj0bpesxg9g
0ukn3wb5t4s6wnootff4zggou9mai2
ix160yh5rmyxim67fmmko20axax7sz
bxqhyoz4g13e4p1b47yps0q3xk05s0
4x11w910q744btpnsx1nqmdtjdtp6u
as5wofip5pta1m0p1xa8nxthhzb655
gxvdmavbg15qmrqcdom1ck6t1548e9
r7u4t4051lab822kfr703r7xtd11pn
giulczmik7swguz0aq8btzirscsul4
2nooznah8ia4y53zft7qi79e057uqb
sydg10rg89qd8vg7aae1xlaclfjg6v
zticsi78rhy5pt1w4wb5uwf6ivqz1m
0fmg75ogx5e2u7oqci1heyx1hscgty
te7amcyp94gcyo0p0kqizpfk77wo0u
u9g5vt4x38df61hv20mk0czkj6h87c
2ly7grbdvjraqtqmvr9gr1fc5meeu7
wsu39mfnhyerp7ib3arj26zfcm9qgh
c9ggj2v7jt1bjtuux4fg61gly2voic

25 ADVANCED CIPHERS - KEYWORD CIPHER

Nearly everyone who's worked with computers knows about passwords. You need them to use most computers, to access some Internet web sites, and to make online purchases.

There's something similar to a password in cryptography. It's called a keyword, and it allows you to encode messages differently when using the same cipher.

As an example of how this is done, we'll choose "SUNNYDAY" as a keyword. The first step is to write down our keyword without repeating any letters. In this case, it means writing just the first "N" and the first "Y."

SUNYDA

The next step is to write down all the letters of the alphabet, in order, skipping any letters in the keyword. The first thing we realize is that we've already written down "A," so we write "B" and continue.

SUNYDABCEFGHIJKLMOPQRTVWXZ

We can then match these letters against the regular alphabet.

Plaintext: ABCDEFGHIJKLMNOPQRSTUVWXYZ

↕↕↕↕↕↕↕↕↕↕↕↕↕↕↕↕↕↕↕↕↕↕↕↕↕↕

Ciphertext: SUNYDABCEFGHIJKLMOPQRTVWXZ

The letters labeled "plaintext" are the ones in your message, and the letters labeled "ciphertext" are the coded letters.

Plaintext	Ciphertext	Plaintext	Ciphertext
A	K	A	S
B	E	B	U
C	Y	C	N
D	W	D	Y
E	O	E	D
F	R	F	A
G	D	G	B
H	A	H	C
I	B	I	E
J	C	J	F
K	F	K	G
L	G	L	H
M	H	M	I
N	I	N	J
O	J	O	K
P	L	P	L
Q	M	Q	M
R	N	R	O
S	P	S	P
T	Q	T	Q
U	S	U	R
V	T	V	T
W	U	W	V
X	V	X	W
Y	X	Y	X
Z	Z	Z	Z

Coding and decoding is often easier if the letters are arranged vertically, as shown on the left. Use of the word, "KEYWORD" as a keyword is shown also on the left.

Our example phrase, "Mary had a little lamb," would be coded as

HKNXAKWKGBQQGOGKHE

using "KEYWORD" as the keyword, and as

ISOXCSYSHEQQHDHSIU

using SUNNYDAY as the keyword.

26 ADVANCED CIPHERS - CAESAR CIPHER WITH KEY

This cipher combines the keyword concept we used in the last chapter with the Caesar cipher.

For our example, we'll choose "JULIUSCAESAR" as a keyword. As before, the first step is to write down our keyword without repeating any letters. In this case, it means writing just the first "U," the first "S," and the first "A."

JULISCAER

The next step is to write down all the letters of the alphabet, in order, skipping any letters in the keyword. The first thing we realize is that we've already written down "A," so we write "B" and continue.

Then, just as in the Caesar cipher, we select a letter shift; for example, six. We can then associate our alphabet letters with the shifted keyword, as shown.

JULISCAERBDFGHKMNOPQTVWXYZ

ABCDEFGHIJKLMNOPQRSTUVWXYZ

AERBDFGHKMNOPQTVWXYZJULISC

"Mary had a little lamb" would be encoded like this.

PAXSHABAOKZZODOAPE

27 ADVANCED CIPHERS - FOUR-SQUARE CIPHER

As we saw in the last chapter, one way to decrease the chances of someone breaking your code is to add more ciphering steps. The Caesar cipher with key combined the keyword cipher and the Caesar cipher, so it should be harder to crack.

The four-square cipher makes things more difficult for a code breaker by using two keywords and operating on pairs of letters at the same time.

There are four charts used to code and decode this cipher, as shown on the next page. Looking at the charts tells you why this is called the four-square cipher.

As for other keyword ciphers, we create an alphabet based on each key using the rules in the previous chapters. For the keywords "SATURDAY" and "SUNDAY" we first create the following keyword alphabets.

```
SATURDYBCEFGHIJKLMNOPQVWXZ
SUNDAYBCEFGHIJKLMOPQRTVWXZ
```

The next step is to put the keyword alphabets into two of the boxes, the red squares in the figure. One trick in doing this is that there are places for only twenty five letters. As is common in many ciphers, we ignore the "J," treating "I" and "J" as the same letter.

The blue squares are filled with the regular alphabet letters, once again not using "J."

Encoding messages is a little harder with the four-square cipher, since we need to do two letters at a time. Once again, we'll use "Mary had a little lamb" as our example message.

The first two letters of our message are "m" and "a." We locate these in the blue squares on the next page like this.

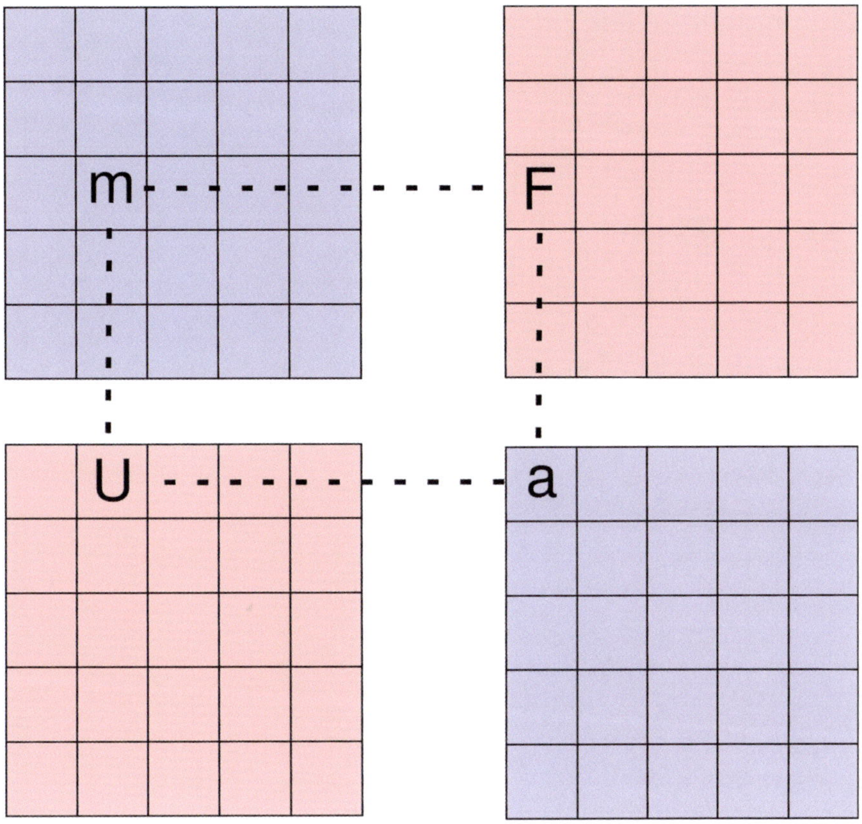

Our plaintext letter pair, "ma," is associated with a ciphertext letter pair, "FU." These are the first two letters of our coded message.

We then do the same with the next letter pair, "ry," to get the coded pair, "OV." Continuing through the entire message, "Mary had a little lamb" would be encoded like this.

FUOVDNSDIYOQKSFSGU

a	b	c	d	e
f	g	h	i	k
l	m	n	o	p
q	r	s	t	u
v	w	x	y	z

S	A	T	U	R
D	Y	B	C	E
F	G	H	I	K
L	M	N	O	P
Q	V	W	X	Z

S	U	N	D	A
Y	B	C	E	F
G	H	I	K	L
M	O	P	Q	R
T	V	W	X	Z

a	b	c	d	e
f	g	h	i	k
l	m	n	o	p
q	r	s	t	u
v	w	x	y	z

28 ADVANCED CIPHERS - POLYBIUS SQUARE

The Polybius square cipher is named after its inventor, the Greek scholar, Polybius, who lived in the second century BC. It's a simpler cipher than the four-square cipher of the previous chapter, so it's easier to code messages using it. That also means that it's easier to crack.

Like other ciphers, the alphabet characters are arranged in an array, as shown in the figure, with the rows and columns numbered. Since it's possible to fit numbers into such an array, they're included also.

	1	2	3	4	5	6
1	A	B	C	D	E	F
2	G	H	I	J	K	L
3	M	N	O	P	Q	R
4	S	T	U	V	W	X
5	Y	Z	0	1	2	3
6	4	5	6	7	8	9

To encode a message, you just write the row and column of each letter. The letter "A" will be encoded as the number 11, the letter "B" will be encoded as the number 12, and the number nine will be encoded as 66.

Our example phrase, "Mary had a little lamb," will be encoded as follows.

31 11 36 51 22 11 14 11 26 23 42 42 26 15 26 11 31 12

29 ADVANCED CIPHERS - VIGENÈRE CIPHER

The Vigenère cipher is important to the history of cryptography, since it uses what's called poly-alphabetic substitution. This means that the substitute characters aren't selected from just one alphabet, but many alphabets. As such, it's a much harder cipher to crack.

The first step in encoding a message using the Vigenère cipher is to choose a key, but unlike the key processing of the ciphers described previously, we make a keyphrase by repeating the key over and over until we get as many letters as are in our message.

For example, if our message is "Mary had a little lamb," and the key is "FLEECESNOW," we would write this.

```
MARYHADALITTLELAMB
FLEECESNOWFLEECESN
```

The table on the next page shows all possible shifted alphabets, somewhat like the Caesar cipher, but these are labeled in rows A-Z. The regular alphabet is shown as the column labels.

To encode the first letter of our message, "M," we look at what letter is in column "M" and row "F." To encode the second letter, we look at what letter is in column "A" and row "L," etc. Our coded message then becomes

```
RLVCJEVNZEYEPINEEO
```

Decoding is done by seeing what letter appears at the top of the column for the cipher letter in the key letter row.

	A	B	C	D	E	F	G	H	I	J	K	L	M	N	O	P	Q	R	S	T	U	V	W	X	Y	Z
A	A	B	C	D	E	F	G	H	I	J	K	L	M	N	O	P	Q	R	S	T	U	V	W	X	Y	Z
B	B	C	D	E	F	G	H	I	J	K	L	M	N	O	P	Q	R	S	T	U	V	W	X	Y	Z	A
C	C	D	E	F	G	H	I	J	K	L	M	N	O	P	Q	R	S	T	U	V	W	X	Y	Z	A	B
D	D	E	F	G	H	I	J	K	L	M	N	O	P	Q	R	S	T	U	V	W	X	Y	Z	A	B	C
E	E	F	G	H	I	J	K	L	M	N	O	P	Q	R	S	T	U	V	W	X	Y	Z	A	B	C	D
F	F	G	H	I	J	K	L	M	N	O	P	Q	R	S	T	U	V	W	X	Y	Z	A	B	C	D	E
G	G	H	I	J	K	L	M	N	O	P	Q	R	S	T	U	V	W	X	Y	Z	A	B	C	D	E	F
H	H	I	J	K	L	M	N	O	P	Q	R	S	T	U	V	W	X	Y	Z	A	B	C	D	E	F	G
I	I	J	K	L	M	N	O	P	Q	R	S	T	U	V	W	X	Y	Z	A	B	C	D	E	F	G	H
J	J	K	L	M	N	O	P	Q	R	S	T	U	V	W	X	Y	Z	A	B	C	D	E	F	G	H	I
K	K	L	M	N	O	P	Q	R	S	T	U	V	W	X	Y	Z	A	B	C	D	E	F	G	H	I	J
L	L	M	N	O	P	Q	R	S	T	U	V	W	X	Y	Z	A	B	C	D	E	F	G	H	I	J	K
M	M	N	O	P	Q	R	S	T	U	V	W	X	Y	Z	A	B	C	D	E	F	G	H	I	J	K	L
N	N	O	P	Q	R	S	T	U	V	W	X	Y	Z	A	B	C	D	E	F	G	H	I	J	K	L	M
O	O	P	Q	R	S	T	U	V	W	X	Y	Z	A	B	C	D	E	F	G	H	I	J	K	L	M	N
P	P	Q	R	S	T	U	V	W	X	Y	Z	A	B	C	D	E	F	G	H	I	J	K	L	M	N	O
Q	Q	R	S	T	U	V	W	X	Y	Z	A	B	C	D	E	F	G	H	I	J	K	L	M	N	O	P
R	R	S	T	U	V	W	X	Y	Z	A	B	C	D	E	F	G	H	I	J	K	L	M	N	O	P	Q
S	S	T	U	V	W	X	Y	Z	A	B	C	D	E	F	G	H	I	J	K	L	M	N	O	P	Q	R
T	T	U	V	W	X	Y	Z	A	B	C	D	E	F	G	H	I	J	K	L	M	N	O	P	Q	R	S
U	U	V	W	X	Y	Z	A	B	C	D	E	F	G	H	I	J	K	L	M	N	O	P	Q	R	S	T
V	V	W	X	Y	Z	A	B	C	D	E	F	G	H	I	J	K	L	M	N	O	P	Q	R	S	T	U
W	W	X	Y	Z	A	B	C	D	E	F	G	H	I	J	K	L	M	N	O	P	Q	R	S	T	U	V
X	X	Y	Z	A	B	C	D	E	F	G	H	I	J	K	L	M	N	O	P	Q	R	S	T	U	V	W
Y	Y	Z	A	B	C	D	E	F	G	H	I	J	K	L	M	N	O	P	Q	R	S	T	U	V	W	X
Z	Z	A	B	C	D	E	F	G	H	I	J	K	L	M	N	O	P	Q	R	S	T	U	V	W	X	Y

30 STEGANOGRAPHY

In a previous chapter, we learned that most computers display images using 16,777,216 colors. It's easy to change each color in an image just slightly, or hide a few patches of a certain color in an image, to encode a secret message. This method of hiding a message is called steganography.

Since numbers are the only language computers understand, colors in computer images are represented using numbers. If we take a photograph and change it so that all the color numbers are made to be an even number slightly higher, or lower, than the actual number, no one would really notice. We could then add our message to the image by making some of the numbers odd.

Although we could just encode our plaintext message that way, we can go through an additional step of using a cipher to encode it first, making the whole process more secure.

It's hard to show such subtle image changes in a book, so the example on the next page shows this concept in a different way. The image appears to be just a mix of circles of random colors with no apparent message.

Our coded message has a key, which is a selection of six specific colors. We can remove all circles, except those of these six colors to reveal our message, the number "1."

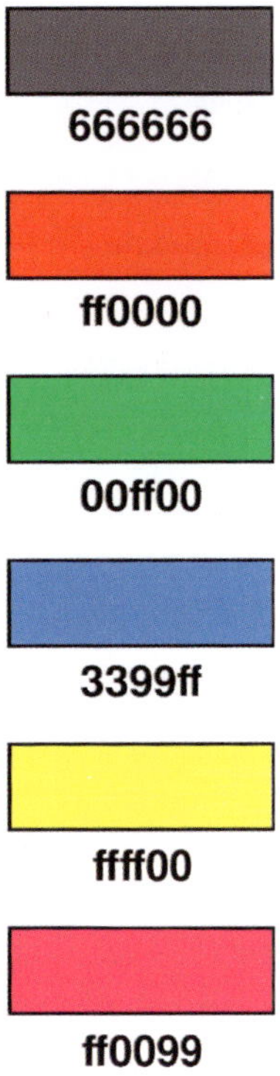

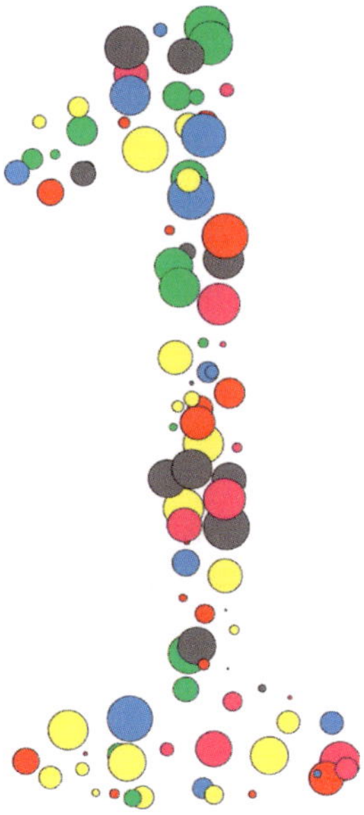

Our secret message is decoded from our steganographic image using the color key, above.

The numbers are the computer codes for these colors. A computer program is able to eliminate all colors except the colors in the key.

31 NUMBER GAMES - HOLEY NUMBERS

Numbers are mostly used to express quantities, such as three apples, or twelve months. Numbers also have other properties that are interesting to mathematicians, like whether they are even, or odd; or, if they can be divided by other numbers. If a number can only be divided by the number "1," or itself, it's called a prime number.

Sometimes, mathematicians think of strange properties of numbers. One of these is the idea of "holey numbers." Holey numbers have holes in them. The digits 0, 4, 6, 8, and 9, contain loops, or holes, and the others don't. Sometimes the number four is written without a loop, so it loses its holey character, as we see in the figure.

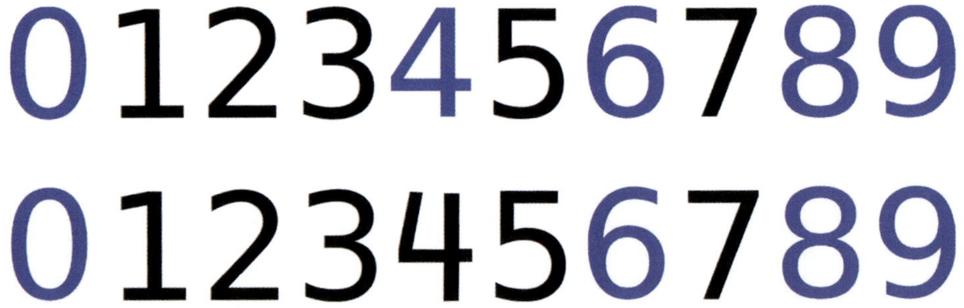

If we write down all numbers containing at least one digit with a hole, we've defined a number sequence. This sequence would start as 0, 4, 6, 8, 9, 10, 14, 16, 18, 19, 20..., and it would continue forever.

We might restrict the numbers to those for which all digits have holes; that is, 0, 4, 6, 8, 9, 40, 44, 46, 48, 49, 60, 64, 66, 68, 69, 80, 84, 86, 88, 89, 90, 94, 96, 98, 99, 400...

You can write your own number sequence. Ask yourself, what holey numbers are also prime numbers?

32 NUMBER GAMES - ADD 'EM UP

Albert Einstein was a famous scientist, and most of his science involved mathematics. There's a story told about young Albert when he was an elementary school student.

His teacher asked the class to add up all the numbers from one to a hundred. The teacher probably needed some free time to grade exams and thought this would keep the students busy for at least a half hour.

Albert almost immediately raised his hand to tell the teacher he had the answer. The teacher, of course, didn't think that was possible, but Albert's answer was correct.

How did Albert do it? He used a concept called symmetry. He noticed that when you arranged the numbers in a certain way, as shown on the next page, all the additions are really easy.

Mathematicians use symmetry to solve a lot of problems that appear at first to be very difficult. When you're doing a problem, see whether symmetry can help in its solution.

	100			100		
1	99	1	+	99	=	100
2	98	2	+	98	=	100
3	97	3	+	97	=	100
4	96	4	+	96	=	100
5	95	5	+	95	=	100
6	94	6	+	94	=	100
7	93	7	+	93	=	100
8	92	8	+	92	=	100
9	91	9	+	91	=	100
~	~	~		~		~
48	52	48	+	52	=	100
49	51	49	+	51	=	100
	50			50		

33 NUMBER GAMES - BASIC NUMEROLOGY

Numbers sometimes seem to have magical properties, like when we use mathematics to make predictions. Primitive people thought that you could use numbers as a means of fortune-telling, a practice called numerology. Modern people are too smart to believe anything like that, although people still talk about "lucky numbers." It's still fun to do numerology as a game.

Numerology can be done for words and names by assigning a number value to each letter. The figure shows the most common way to do this.

1	2	3	4	5	6	7	8	9
A	B	C	D	E	F	G	H	I
J	K	L	M	N	O	P	Q	R
S	T	U	V	W	X	Y	Z	

According to the table, the letters A, J and S would have the value 1, the letters B, K and T would have the value 2, etc.

To get a single digit from a word, you need to do some addition. Taking the name, "Mary," as an example, here's how it's done.

$$
\begin{aligned}
M &= 4\\
A &= 1\\
R &= 9\\
Y &= 7\\
\hline
SUM &= 21
\end{aligned}
$$

$$2 + 1 = 3$$

You can see the steps involved to change the word to a single digit number.

1) Get the number value for each letter of the word from the chart.
2) Add all the numbers.
3) If this sum is greater than a single digit, add the digits of the sum.
4) If this sum is still greater than a single digit, add the digits again.
5) Continue until a single digit is reached.

We can even do this for long phrases. The phrase, "Mary had a little lamb," adds up to 69, so 6 + 9 = 15, and 1 + 5 = 6.

34 NUMBER GAMES – NUMEROLOGY WORD ART

It's possible to make art from words, names and phrases. Here are two simple methods to make a line drawing from a word.

The first example uses the graphing technique of plotting lines on a grid. This is a common technique for representing numbers, like temperature change from day to day. The way that we'll use the grid is as a "phase space," which is actually easy to do.

We mark the sides of the grid with the alphabet characters a-z. We can now plot our lines as shown in the figure for the word, "Wednesday," using the following method.

1) Locate the first letter **w** on the horizontal axis and the second letter **e** on the vertical axis. Place a point at that location.
2) Locate the second letter **e** on the horizontal axis and the third letter **d** on the vertical axis. Place a point at that location.
3) Draw a line between the points.
4) Locate the third letter **d** on the horizontal axis and the fourth letter **n** on the vertical axis. Place a point at that location.
5) Draw a line between the last two points.
6) Continue through all the letters of the word or phrase.

At the end, you can remove the grid to just show the lines. Your name will look different than the names of your friends, and you can use these lines as your "signature."

There's another method to do this that's a little easier. This method puts the alphabet letters into an array, much like the Polybius cipher in a previous chapter.

In this case, as in many ciphers, the letters "I" and "J" share the same box. Using the template on the next page, you just draw lines between the letters in the order they appear in the word or phrase.

You can see some examples of word art on the last page of this chapter.

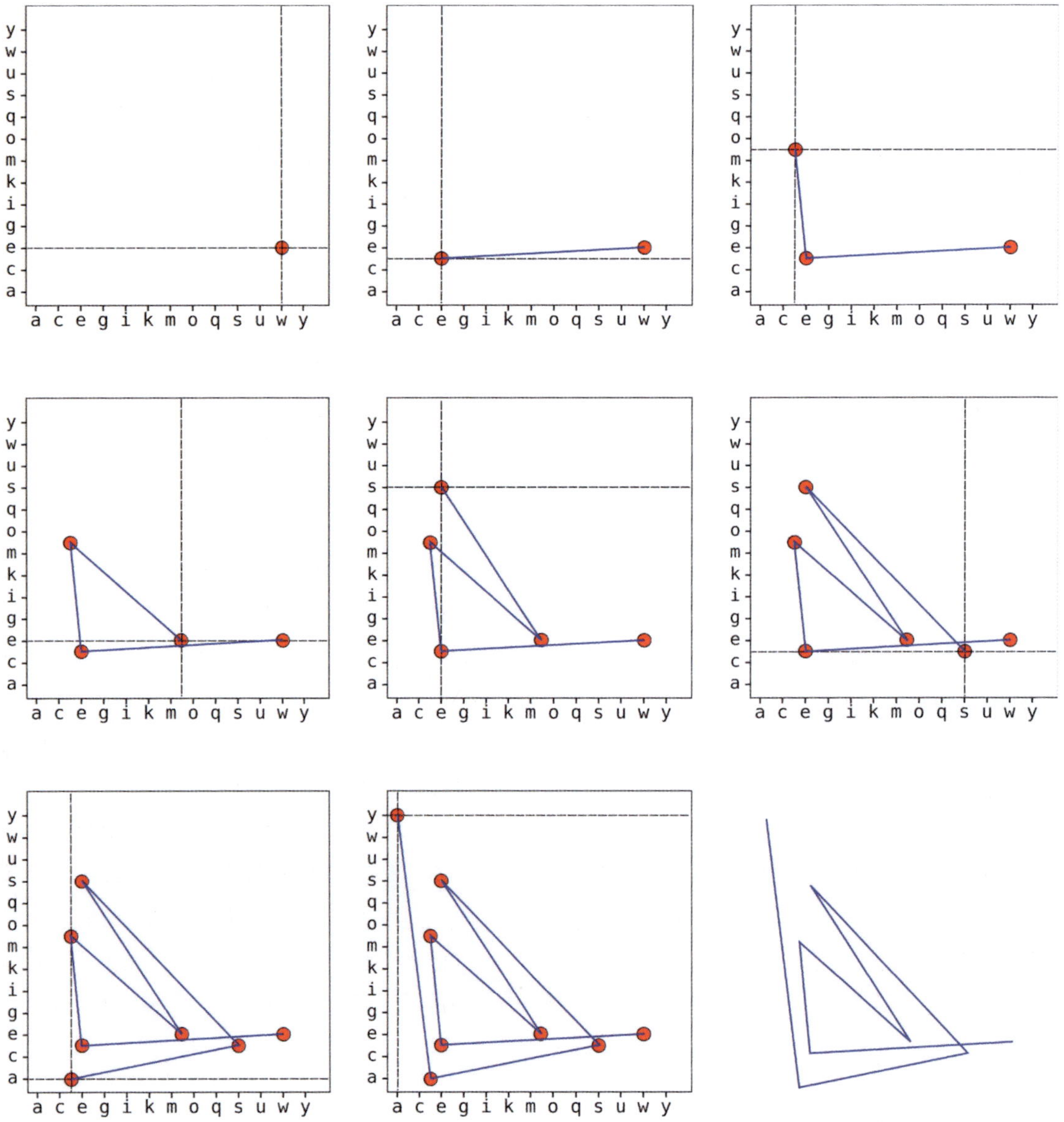

A	B	C	D	E
F	G	H	IJ	K
L	M	N	O	P
Q	R	S	T	U
V	W	X	Y	Z

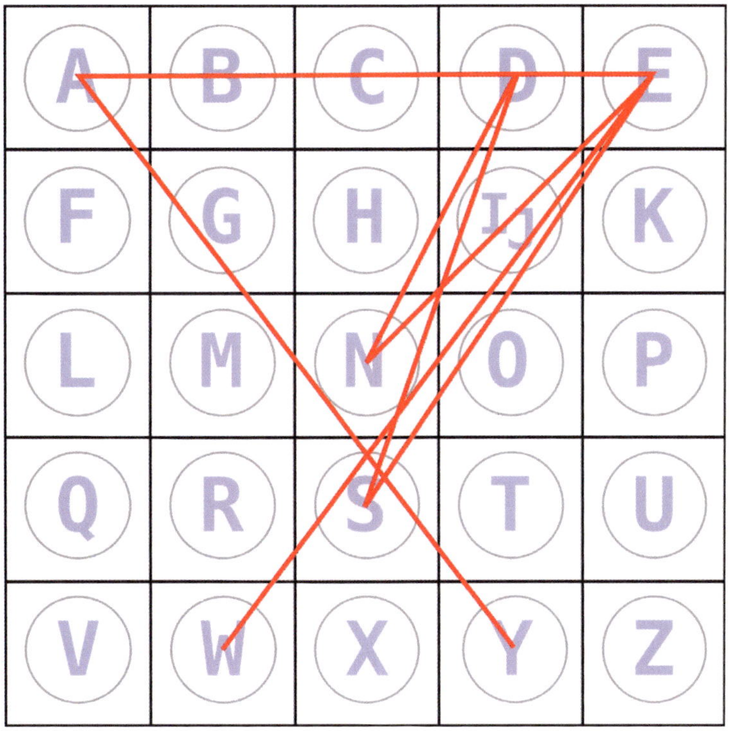

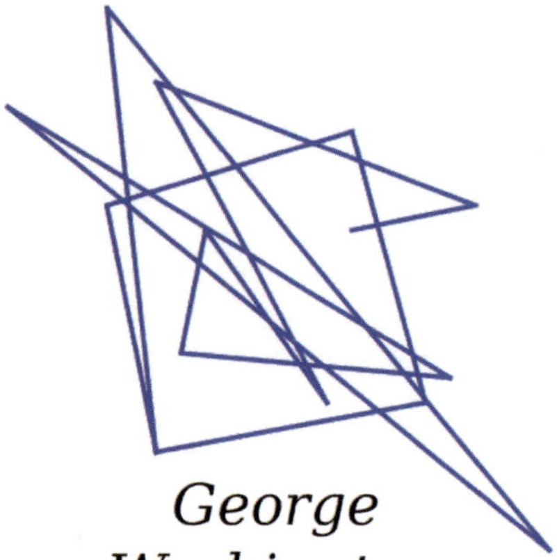

George Washington

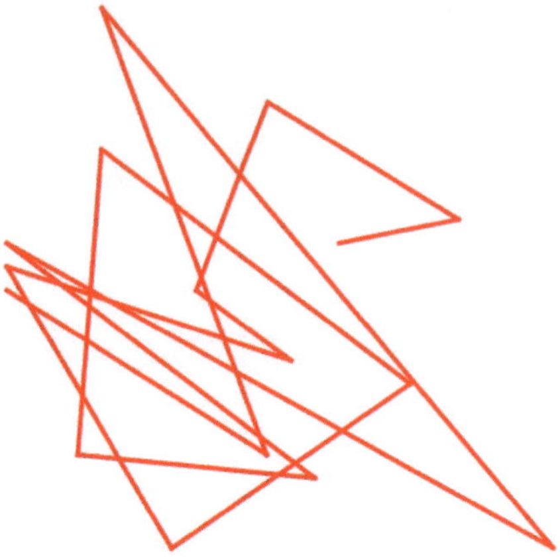

Alexander Hamilton

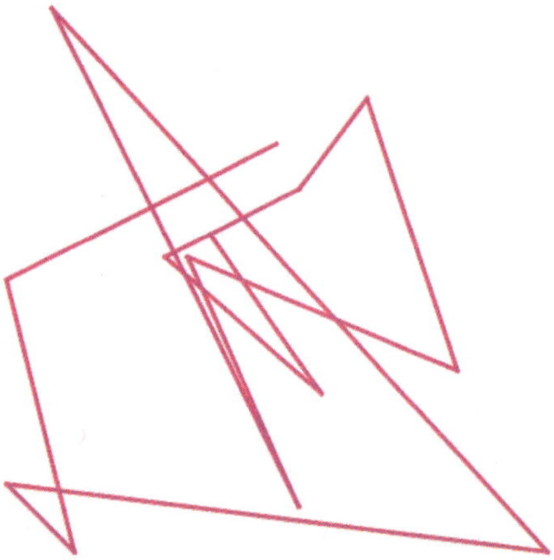

John Quincy Adams

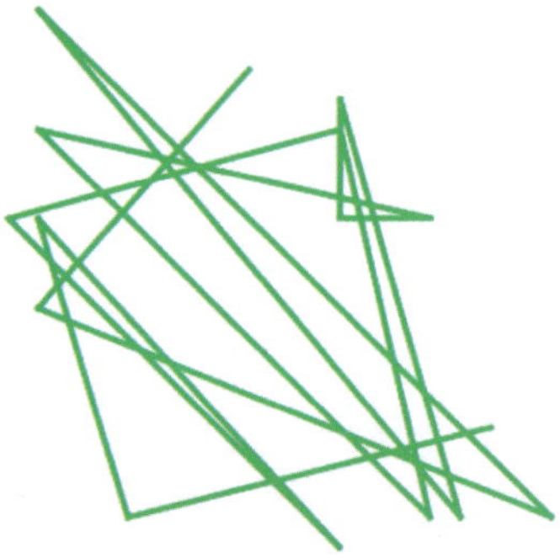

Theodore Roosevelt

Scrambled Eggs

Cheese Pizza

35 NUMBER GAMES - MAGIC SQUARES

A magic square is a grid filled with numbers. Just like numerology, there's nothing really "magic" about a magic square. Magic squares have that name because they have a special property.

In a magic square, the numbers in each row add up to a certain number. Not only that, but the numbers in each column add up to the same number. But wait, there's more! Both sets of diagonal numbers also add up to the same number.

Simple magic squares were made as early as 650 BC. Some were thought important enough to be inscribed on the walls of temples.

Today, computers make it easy to make magic squares, and there are mathematical formulas that allow people to do smaller ones by hand, or with a calculator.

There's not much fun in just pushing a button and getting a magic square. What's more fun is to try to make your own. A good place to start is a 3-by-3 square.

The examples on the next page are two five-by-five magic squares. The first square has the "magic" number, the number that the rows, columns and diagonals add up to, 65. The second square also has rows and columns, but only one diagonal, that add to 65.

15	8	1	24	17
16	14	7	5	23
22	20	13	6	4
3	21	19	12	10
9	2	25	18	11

1	24	17	15	8
7	5	23	16	14
13	6	4	22	20
19	12	10	3	21
25	18	11	9	2

36 ADVANCED TOPICS - A SIMPLE BLOCK CIPHER

As we learned in the chapter about the four-square cipher, ciphers that operate on more than one letter at a time are harder to crack. Ciphers that operate on many letters at once are called block ciphers.

In a block cipher, the block is the number of characters you process at one time. The processing you do can involve both substitution and permutation. In the following example, we just use permutation. In this cipher we'll generate a key from the phrase, "Fool me once, shame on you," using our usual technique.

FOLMENCSHAYUBDGIJKPQRTVWXZ

Our message will be Lincoln's Gettysburg Address, "Four score and seven years ago, our fathers..." We'll operate on blocks that are twenty six letters long, so our first block will be

FOURSCOREANDSEVENYEARSAGOO

Our process is as follows.

1) Write the normal alphabet of twenty six letters above our message block.
2) Find the first two letters of our key in the regular alphabet. These are **F** and **O**.
3) Swap the two letters of our message below these letters. These are the letters **C** and **V**.

FOURSVOREANDSECENYEARSAGOO

4) Find the next two letters of our key in the regular alphabet. These are **L** and **M**.
5) Swap the two letters of our message below these letters. These are the letters **D** and **S**.

FOURSVOREANSDECENYEARSAGOO

6) Continue until all the letter pairs of the key have been used.

Our end result is as follows.

RREOEVEFONASDSCNEAUYOASORG

The figure on the next page shows the step-by-step process for this block cipher.

Passphrase: F O O L M E O N C E S H A M E O N Y O U

Key: F O L M E N C S H A Y U B D G I J K P Q R T V W X Z

Plaintext: F O U R S C O R E A N D S E V E N Y E A R S A G O O   U R F A T H

Swap F-O:
A B C D E **F** G H I J K L M N **O** P Q R S T U V W X Y Z
F O U R S V O R E A N D S E C E N Y E A R S A G O O

Swap L-M:
A B C D E F G H I J K **L** **M** N O P Q R S T U V W X Y Z
F O U R S V O R E A N S D E C E N Y E A R S A G O O

Swap E-N:
A B C D **E** F G H I J K L M **N** O P Q R S T U V W X Y Z
F O U R E V O R E A N S D S C E N Y E A R S A G O O

Swap C-S:
A B **C** D E F G H I J K L M N O P Q R **S** T U V W X Y Z
F O E R E V O R E A N S D S C E N Y U A R S A G O O

Swap H-A:
A B C D E F G **H** I J K L M N O P Q R S T U V W X Y Z
R O E R E V O F E A N S D S C E N Y U A R S A G O O

Swap Y-U:
A B C D E F G H I J K L M N O P Q R S T **U** V W X **Y** Z
R O E R E V O F E A N S D S C E N Y U A O S A G R O

Swap B-D:
A **B** C **D** E F G H I J K L M N O P Q R S T U V W X Y Z
R R E O E V O F E A N S D S C E N Y U A O S A G R O

Swap G-I:
A B C D E F **G** H **I** J K L M N O P Q R S T U V W X Y Z
R R E O E V E F O A N S D S C E N Y U A O S A G R O

Swap J-K:
A B C D E F G H I **J** **K** L M N O P Q R S T U V W X Y Z
R R E O E V E F O N A S D S C E N Y U A O S A G R O

Swap P-Q:
A B C D E F G H I J K L M N O **P** **Q** R S T U V W X Y Z
R R E O E V E F O N A S D S C N E Y U A O S A G R O

Swap R-T:
A B C D E F G H I J K L M N O P Q **R** S **T** U V W X Y Z
R R E O E V E F O N A S D S C N E A U Y O S A G R O

Swap V-W:
A B C D E F G H I J K L M N O P Q R S T U **V** **W** X Y Z
R R E O E V E F O N A S D S C N E A U Y O A S G R O

Swap X-Z:
A B C D E F G H I J K L M N O P Q R S T U V W **X** Y **Z**
R R E O E V E F O N A S D S C N E A U Y O A S O R G

37 ADVANCED TOPICS - LETTER FREQUENCY ANALYSIS

Some letters are used frequently, and some letters are seldom used. It will come as no surprise that the last letter of the alphabet, "z," is almost never used. Just examine a dictionary or encyclopedia to see how few pages are devoted to "z-words."

How frequently letters are used in words is one way that cryptographers can crack secret codes. This cracking method is called letter frequency analysis.

The table and graph on the next page show, on average, how often English letters are used. The most frequently used letter is "e." More than one out of every ten letters written is the letter "e," and one letter "t" appears, on average, for every eleven letters.

If we used a cipher, such as the Caesar cipher, which substitutes a "g" for every "e," and a "v" for every "t," then the letters "g" and "v" would occur more than other letters. A good cryptographer would know he or she should substitute "e" and "t" for the letters "g" and "v" in the message. Then words like "the" will start to appear.

Letter frequency is only approximate for any message, as the table of letter frequency for Arthur Conan Doyle's book, "The Adventures of Sherlock Holmes," shows. Still, it allows critical code-breaking clues in many cases.

English Language

Letter frequency (%)

e	12.70	m	2.41
t	9.06	w	2.37
a	8.17	f	2.23
o	7.51	g	2.02
i	6.97	y	1.97
n	6.75	p	1.93
s	6.33	b	1.49
h	6.09	v	1.04
r	5.99	k	0.75
d	4.25	j	0.15
l	4.03	x	0.15
c	2.78	q	0.10
u	2.76	z	0.07

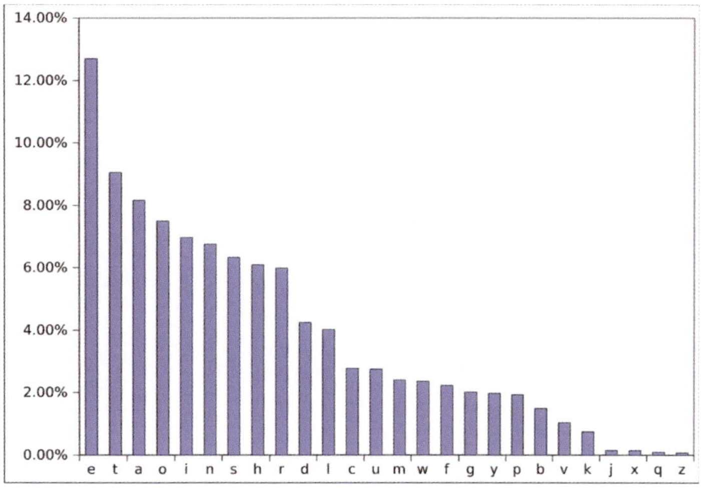

Sherlock Holmes

Letter frequency (%)

e	12.30	w	2.61
t	9.03	c	2.43
a	8.14	y	2.18
o	7.76	f	2.08
i	6.98	g	1.83
h	6.73	p	1.58
n	6.64	b	1.47
s	6.30	v	1.03
r	5.68	k	0.82
d	4.30	x	0.13
i	3.97	j	0.10
u	3.04	q	0.10
m	2.73	z	0.03

38 ADVANCED TOPICS - LOADED DICE

Simple matters are sometimes settled by a coin flip, and dice are used in some board games as a way to determine how a player should move. The assumption in these cases is that the coin is “fair,” and the dice aren't loaded.

A fair coin is one that shows an equal number of heads and tails after being flipped many times. This confirms our idea that the coin has an equal chance of landing one way, or the other.

The “load” in the concept of loaded dice was usually a small lead weight inserted into one of the die faces and then hidden. The idea was to make the dice fall on that face more often, to one player's advantage.

Computers make it easy to cast loaded dice, as shown in the figure on the next page. In that figure, you see also that a hundred tosses of a fair die does not give equal probability for each face. Things will average out better if we cast a thousand dice, but things involving chance are never precise.

Although computers can throw dice and give results that look very much like real dice throws, they aren't. Computers don't generate random numbers, they generate pseudorandom numbers. If we know the program the computer uses to generate these numbers, we can predict what the numbers will be.

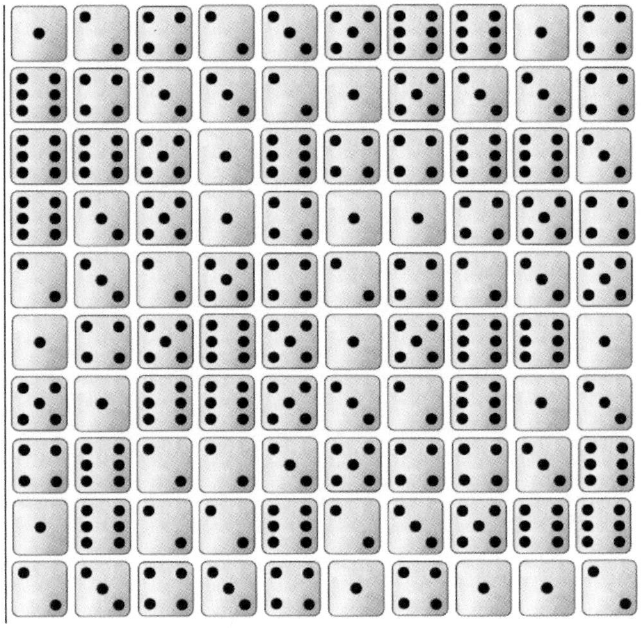

Fair Dice

Die face 1 (%) = 16

Die face 2 (%) = 15

Die face 3 (%) = 16

Die face 4 (%) = 18

Die face 5 (%) = 14

Die face 6 (%) = 21

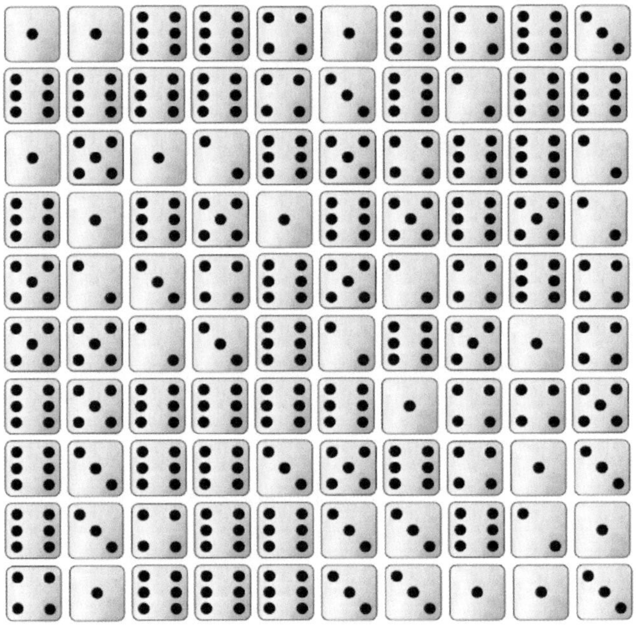

Loaded Dice (6 favored)

Die face 1 (%) = 14

Die face 2 (%) = 9

Die face 3 (%) = 13

Die face 4 (%) = 13

Die face 5 (%) = 13

Die face 6 (%) = 38

39 ADVANCED TOPICS – FAUXCRYPT MOCK CIPHER

Most ciphers are designed to hide messages. Here's one that's designed to be amusing. It's a permutation cipher with a twist. All permutations are done within single words, and they're designed in such a way that the resulting cipher can be understood by people, but not necessarily by computers

The inspiration for this cipher was an anonymous posting on the Internet in September, 2003, purportedly from Cambridge University. The posting reads as follows:

"Aoccdrnig to a rscheearch at an Elingsh uinervtisy, it deosn't mttaer in waht oredr the ltteers in a wrod are, the olny iprmoetnt tihng is taht frist and lsat ltteer is at the rghit pclae. The rset can be a toatl mses and you can sitll raed it wouthit porbelm. Tihs is bcuseae we do not raed ervey lteter by it slef but the wrod as a wlohe."

This statement is still unattributed, but whatever its source, it makes the point that text can be garbled but still readable. The author has written a cipher, called FauxCrypt, that does this scrambling in a way that maintains reasonable readability.

Here are some of the things that FauxCrypt does:

1) Vowels that occur in pairs (digraphs) are swapped. For example, “oe” become “eo.”
2) The order of vowels and consonants in a word are kept the same, but some vowels are shifted forwards or backwards.
3) Some consonants, called “risers,” extend above other characters. The risers are b, d, f, h, k, l and t. Other consonants, called “danglers,” extend below the line of text. The danglers are g, j, p, q and y. Text is generally readable when a "riser" and a "dangler" are swapped; e.g., dg becomes gd).

The following table lists English digraphs and their frequency. The following paragraph is a text processed using FauxCrypt.

Common English language digraphs and their frequency of occurrence.

Digraph	Frequency (%)	Digraph	Frequency (%)	Digraph	Frequency (%)
nt	0.56	hi	0.46	de	0.09
ha	0.56	is	0.46	se	0.08
es	0.56	or	0.43	le	0.08
st	0.55	ti	0.34	sa	0.06
en	0.55	as	0.33	si	0.05
ed	0.53	te	0.27	ar	0.04
to	0.52	et	0.19	ve	0.04
it	0.5	ng	0.18	ra	0.04
ou	0.5	of	0.16	ld	0.02
ea	0.47	al	0.09	ur	0.02

the lnadlord had there dauthgres, who lookde at the goose wiht envoius eyse. tehy had neevr sene such a wnoedrful bird, and lnodge to haev at laest one of its faehtres. "ah," thuohgt the eledst, "i sahll sono haev an optotrunipy to pluck one of tehm;" and so it hapdnepe, for not lnog aftre the yuong man left the room. she instnatly wnet up to the bird and took hold of its wnig, but as she did so, the fnigre and thumb remanide and stuck fsat. in a shotr time aftre the secnod ssitre came in wiht the full extectpaoin of ganinig a goledn faehtre, but as she tuocehd her ssitre to moev her from the bird, her hnad stuck fsat to her ssitre's derss, and neihtre of tehm cuold fere hreself. at lsat, in came the third ssitre wiht the same itnneitno. "keep away, keep away!" sceramde the ohtre two; "in haeevn's name keep away!"

Excerpt from The Golden Goose, from Grimm's Fairy Tales. This excerpt is from The Project Gutenberg eBook, Fairy Tales Every Child Should Know, Edited by Hamilton Wright Mabie.

40 INVENT YOUR OWN CIPHER

As you've learned from the previous chapters, the two basic principles of any cipher are permutation and substitution. Each cipher does one, or both, in its own way, sometimes using keywords to define the process.

See whether you can apply what you've learned to create your own unique cipher. One important thing is that you need more than just a method to scramble your message. You need a reliable way to unscramble it, also!

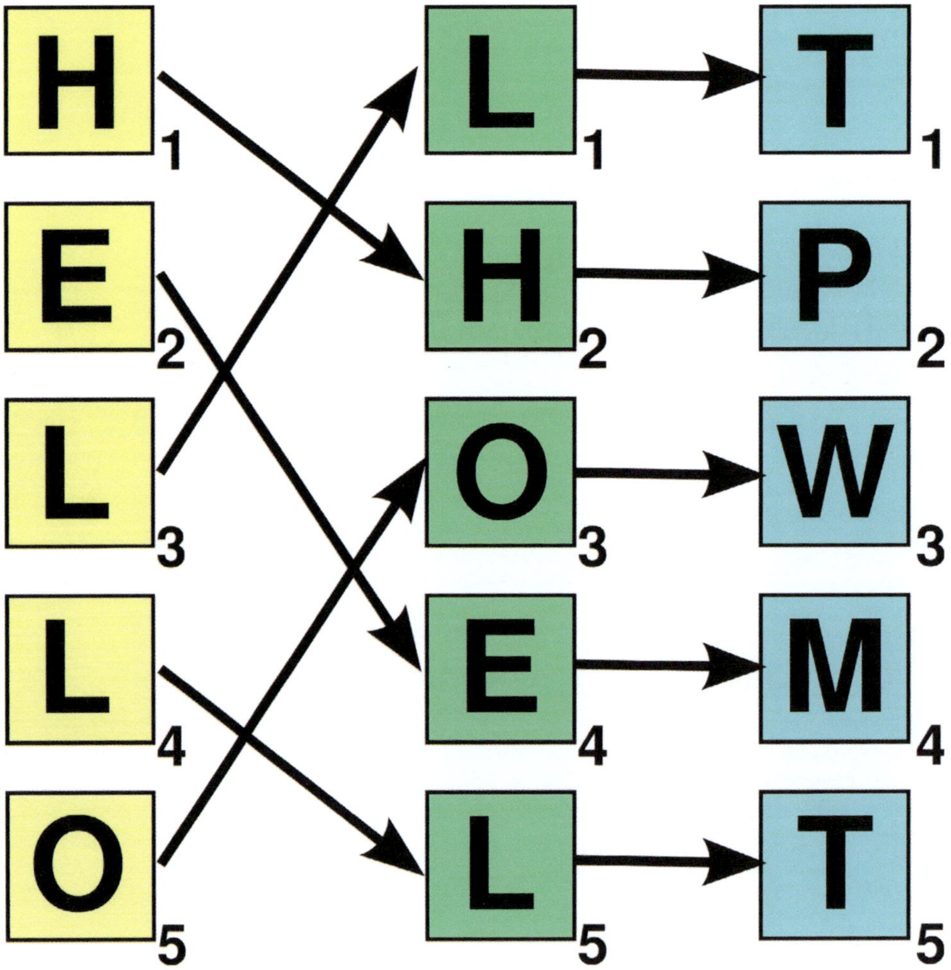

ABOUT THE AUTHOR

Dev Gualtieri received his PhD in 1974 and had a thirty-five year research career in physics and materials science.

Dr. Gualtieri is listed as an inventor on more than thirty US patents, and on numerous international patents. He has published many scientific papers in the areas of superconductivity, chemical thermodynamics, magnetism, electronics and computer science. At one time, he was an internationally recognized expert in crystal growth.

Dr. Gualtieri is now retired, and he resides in Northern New Jersey with his wife, Anne. They have a son and daughter who reside with their spouses in Pennsylvania.

Printed in Dunstable, United Kingdom

91342779R00051